D0674158

CHINESE GRAIN ECONOMY AND POLICY

Dedicated to the memory of
my loving Mother, Weng Tang Song

Chen Liang Yu

Chinese Grain Economy and Policy

by

Chen Liang Yu

London Export Corp. Ltd.
Beijing Office

and

Allan Buckwell

Department of Agricultural Economics
Wye College
Ashford
Kent TN25 5AH
UK

C·A·B INTERNATIONAL

C·A·B International
Wallingford
Oxon OX10 8DE
UK

Tel: Wallingford (0491) 32111
Telex: 847964 (COMAGG G)
Telecom Gold/Dialcom: 84: CAU001
Fax: (0491) 33508

British Library Cataloguing in Publication Data
Yu, Chen Liang
Chinese grain economy and policy.
1. China. Cereals. Production
I. Title II. Buckwell, Allan E.
338.1731095

ISBN 0-85198-642-0

Printed in the UK by Redwood Press Ltd., Melksham
Bound in the UK by Green Street Bindery, Oxford

Contents

Preface

The ideas reported in this book were hatched over the period 1980 to 1988 whilst the senior author was working first in the Agricultural Bureau of the State Planning Commission in Beijing and then at Wye College, London University. This was an exciting period to be studying Chinese grain policy because it appeared that the lessons of the past were being heeded and organizational and pricing policies put in place which promised to lift the grain economy once and for all onto a higher plane of productivity and reliability. The erratic changes in policy which had previously caused so much hardship seemed to be a thing of the past. The liberalization which had taken place since 1978, whilst far from complete, changed the nature of incentives faced by grain producers. The prima facie evidence was that this was an important factor in the improved performance of the grain sector. A major task for economists was therefore to study carefully the sequence of events throughout the tumultuous developments since the formation of the People's Republic and to try and disentangle cause and effect. These were precisely the objectives of this study.

The events of 1989 throughout the Communist world add a new dimension of urgency to the need to understand the basic economic and technological relationships of the agricultural economy. The record of underachievement of the socialist economies in the production in consumer goods and especially even the most basic foodstuffs demands explanation. It was hoped that the combination of application of elementary analytical ideas of Western agricultural economics with a detailed knowledge of the convolutions of Chinese grain policy and institutions might lay the foundations for this explanation.

A further reason for this work was that for many years China's rural development was held up as a successful example for other developing countries to follow. It therefore seems reasonable that the model should be subjected to scrutiny so that the links between policy and performance could be better understood.

The empirical work described here covers the period 1952 to the late 1980s. At the time of writing (early 1990) all policy, including grain policy, was in a period of great uncertainty. The rapid liberalization had been faltering during the late 1980s and was halted during 1989, but the new lines of policy were unclear. There seemed a very real danger that the rural economy might be faced with the repetition of the policy reversals in the past which had been so damaging. It is hoped that this book may contribute towards avoiding this outcome.

We would like to thank our numerous colleagues for their assistance in the preparation of this book. In particular, in China we benefited from the help and advice of Mr Liu Zhongyi, Senior Economist and Vice Minister of the State Planning Commission (SPC); Mr Liu Yaochuan, Senior Economist and Head of the Bureau of Agriculture, SPC; Mrs Li Li, Economist and Head of the Comprehensive Planning Branch, Bureau of Agriculture, SPC; Mr Huan Yaohui, Director of the Planning Bureau, Ministry of Agriculture, Animal Husbandry and Fisheries (MAAF); Mr Zhang Tong, Chief Agricultural Economist of the MAAF; Prof An Xiji, Beijing University of Agriculture; Professor Zhang Xiangshu, The People's University of China; and Mr Zhou Zhilin, Deputy Manager of China's Statistical Information and Consultancy Service Centre, State Statistical Bureau. We would also like to record our thanks to the numerous officials and academics we met during our visits in May 1988 to Beijing, Xian, Nanjing, Hanzhou and Fuzhou.

In Britain we have benefited from the comments of Professors George Peters of Oxford University, Gordon White of the Institute for Development Studies, Sussex University and the late Professor Kenneth Walker of the School of Oriental and African Studies, University of London. We are also grateful to Dr Bruce Stone of the International Food Policy Research Institute who in his review of the manuscript made numerous perceptive and constructive suggestions. We are indebted to all these scholars but of course the text presented is entirely the responsibility of the authors.

Throughout the course of this study we have been assisted in a practical way by the Committee of Vice Chancellors and Principals of the Universities of the United Kingdom and the Central Research Fund Committee of the University of London. We gratefully acknowledge the financial assistance of the British Council, the Ministry of Agriculture, Animal Husbandry and Fisheries of China and especially the United Kingdom Overseas Development Administration.

In acknowledging the generous support of the above organizations, in particular the Overseas Development Administration, we emphasize that the views expressed in this book are our own and not those of these organizations.

Finally, we owe a great deal to Lindy Millbank and Anne Curtis who patiently produced a manuscript from a most unpromising draft!

Abbreviations

Institutions

AAPCs = Advanced Agricultural Producers' Co-operatives
APH = Agricultural Publishing House
ATECs = Agrotechnical Extension Centres
CAAS = Chinese Academy of Agricultural Science
CCP = The Chinese Communist Party
EAPCs = Elementary Agricultural Producers' Co-operatives
FAO = The Food and Agricultural Organisation
ITRI = International Trade Research Institute of MOFERT (see below)
MAAF = Ministry of Agriculture, Animal Husbandry and Fishery
MATs = Mutual Aid Teams
MOFERT = Ministry of Foreign Economic Relations and Trade
NAZC = National Agricultural Zoning Committee
PRC = The People's Republic of China
PRS = The Production Responsibility System
SPC = The State Planning Commission
SSB = The State Statistical Bureau
WB = The World Bank

Statistical Sources

JZNQ = Jianou Xian Zonghe Nongye Quhua 1984 (Comprehensive Agricultural Zoning of Jianou County 1984)
NCTJNJ = Zhongguo Nongcun Tongji Nianjian (Chinese Rural Statistical Yearbook)
NJZL = Nongye Jingji Ziliao (Statistics of Agricultural Economy)
PNQ = Pindu Xian Nongye Quhua 1984 (Report on Agricultural Zoning of Pindu County 1984)
TJNJ = Zhongguo Tongi Nianjian (Chinese Statistical Yearbook)
ZCWTZ = Zhongguo Caimao Wujia Tongji Ziliao (China's Statistical Data on Trade and Prices)

Introduction

China is a large country still heavily based on agriculture. According to the *Chinese Statistical Yearbook 1985*, 68% of the Chinese population lives in the countryside, 74% of the total labour force is involved in the agricultural sector, and 29% of GNP was contributed by the agricultural sector in 1984. The development of agriculture is clearly the foundation for the development of the Chinese national economy.

China is the largest grain producer in the world. In the record year of 1984, 407 million tonnes of grain were produced, this sustains the lives of one billion Chinese and is potentially an important factor in the stabilization of the world grain market. By its sheer size, China has the capability, as yet not examined to any great extent, of significantly affecting world grain markets.

The historical development of Chinese agriculture is highly influenced by the political environment. Since the founding of the People's Republic of China (PRC), the government has tightly controlled or deeply intervened in agriculture. Government policy directly influenced rural institutions and production organization. Thus the decision-making and management processes concerning what and how to produce, the pricing and marketing system and the technological improvement of the production system were all under close governmental control. Erratic changes in policies and institutions help explain the zig-zag course of Chinese grain and agricultural production during the last three decades.

OBJECTIVES. It is not surprising, given the importance of the agricultural economy in the development of the whole national economy and its worldwide impact, that much research has been done on this sector by Chinese and Western scientists. This effort has intensified since the important changes in policy in 1978 and much has been concerned with assessing the impact of the new policies on agricultural development.

However, despite these efforts, a set of key issues require further analysis:

(a) What are the main factors which determine Chinese peasants' supply response behaviour? What rôle have price signals played in the supply response behaviour of producers under the highly controlled economic system?

(b) Is it possible to identify systematic policy patterns within the apparently bewildering series of changes in policy and which influenced the historical development of Chinese agriculture? If so, what are these patterns, what were the policy mechanisms, how do they work and what are the impacts on grain as well as agricultural production more generally?

(c) What are the main factors contributing to the demand for grain in China under the planned economic system? What are the relationships between the grain production, domestic consumption and foreign trade and how does this impact on the world market?

These issues are vital for a better understanding of the nature of the new policies and their potential impact on domestic grain production, consumption and trade towards the end of the twentieth century.

CONTENTS. The book is divided into three parts. The first sets out in Chapters One to Three the background to Chinese agriculture and policy and develops a framework for addressing the questions outlined above. The second part contains in Chapters Four to Seven the core analysis of production, consumption and trade in grain over the last three decades. The final part comprising Chapters Eight to Ten, looks ahead, projecting the possible future course of the grain economy and discussing the implications of likely developments.

In more detail, Chapter One contains a general review of the historical development of China's agriculture. This covers the basic features of the natural resource endowment, the improvement in production conditions, and the trends in agricultural production and food consumption. Chapter Two reviews the key policy issues which influenced the historical development of the grain economy as well as the agricultural economy more generally. This leads in Chapter Three to a series of hypotheses about the policy patterns and mechanisms pursued by China over three decades of development. In turn, those questions suggest the analytical approach which is described in Chapter Three and implemented in Part II.

The four chapters of Part II contain the main empirical analysis of the historical development of the grain sector and provide answers to the three questions posed above. In Chapter Four, a theoretical explanation for grain supply under China's circumstances is specified. Time-series information relating to China as a whole and also to two regions is then used to estimate statistical relationships. The theoretical hypothesis under test is that during the last three decades since the founding of the PRC, the complexity of policy can be summarized in just two alternative policy patterns which can explain the zig-zag course of China's grain as well as agricultural development. Having tested this major hypothesis with the empirical evidence in Chapter Four, the following chapter explores the policy issues in more detail focussing on peasant farmers' supply response behaviour, the policy patterns, their mechanisms and impacts on the historical development of grain and agricultural production. Chapter Six switches attention from production to domestic grain consumption, in which the four categories of food, feed, seed and industrial use of grain products are considered. In the specification and estimation of the consumption relationship for food grain, the theoretical hypothesis tested is that under the Chinese circumstances, consumption of food grain was not only dependent upon income and price levels, but also, and perhaps more importantly, on the available production and the growth in population. Study of the nation's grain trade is conducted in Chapter Seven. The complexity of China's grain trade, the motivations for this trade and the trade patterns adopted during the different development periods are discussed. This is followed by the statistical analysis of the trade flows of rice exports, wheat imports and net trade of grain products. The hypothesis tested is that apart from the price signals in the world market, both the key factors determining China's grain trade are the domestic circumstances. This trade is a residual consideration rather than a primary concern. The empirical analysis in these four chapters of Part II broadly support the specified hypotheses. These answers to the three key policy issues raised above are found in the results of statistical analysis combined with rather more qualitative evidence.

Part III comprises two chapters which are concerned with projecting the future course of the grain sector and the outlook for Chinese grain trade and a final chapter drawing together the lessons of the last thirty years of grain policy. Chapter Eight considers the prospects for the future development of China's grain economy to the end of the century. With the identification of the likely general social and economic climate to be faced in the foreseeable future and detailed assumptions made about the relevant policy issues, three alternative scenarios are specified in projecting the future grain production and consumption. The results of the historical analysis are assessed for their usefulness in making predictions. Where no fundamental structural change is expected, the statistical models are used for prediction. Otherwise, hypothetical projections are made. The prospects for future trade discussed in Chapter Nine start with the calculation of grain balances. This enables investigation of the net trade position, the scale of trading and its composition. The final step is to consider the possible impacts of changes in the Chinese net trade position on the world market. The final chapter reviews the failures of the past andanticipates the opportunities and challenges of the future in continuing to feed a quarter of the world's population with only a sixteenth of its arable land.

Part I

Background and Hypotheses

Chapter One

Basic Features of Chinese Agricultural Development

The prime objective of this book is to examine in detail the impact of policy on China's grain economy. Before this can be done, it is first necessary to summarize the basic features of Chinese agriculture. This will cover the natural resource endowment, the improvement of production conditions and the trends in agricultural production and food consumption. Because of the inaccessibility of data to non-Chinese scholars, the last section of this chapter summarizes and assesses the data available.

1.1 Natural Resource Endowment

1.1.1 General features

China has 9.6 million sq.km. of territory. She ranks third in size in the world and her geographic and climatic richness is unequalled by any other country. These varied conditions have required the Chinese to adopt equally varied systems of crop farming, animal husbandry, forestry, fishery and sideline occupations.

China's terrain varies from mountainous areas, with elevations of more than 4,000 metres in the west to lowlands in the east. Of the total land area, mountains make up 33%, plateaux 26%, basins 19%, plains 12% and hilly lands 10%. Thus the upland areas, which including mountain regions, hilly land and comparatively rugged plateaux, cover more than two thirds of the total area. These terrains intermingled and combined with the effects of latitude and climate have created a complex pattern of natural regions.

There are four large plateaux in China, the Qinghai-Xizang Plateau, the Nei Mongol (Inner Mongolia) Plateau, the Loess Plateau and the Yunnan-Guizhou Plateau. On the first two there are broad natural grasslands where herdsmen of many nationalities have grazed their livestock since ancient times. The Loess Plateau comprises deep wind borne soil which is suitable for developing forestry, arable farming and animal husbandry. The Yunnan-Guizhou Plateau contains numerous small basins in which farming can be pursued.

The Sichuan Basin, the Talimu Basin, the Zhunger Basin and the Chaidamu Basin are the four large basins in China. Among them, the Sichuan Basin is referred to as "the land of plenty" due to its warm and humid climate and its abundance of agricultural produce.

The Dongbei (northeast China) Plain, the Huabei (north China) Plain and the plain along the middle and lower reaches of the Changjiang (Yangtze) River are the three large plains in China. The Dongbei Plain supports large crops of soybeans, sorghum and maize given its smooth terrain and fertile soil. The Huabei Plain accounts for over one fifth of the total cultivated area of the country and is an important region for the production of grain, cotton, vegetable oils and tobacco. The plain along the middle and lower reaches of the Changjiang River is China's famous "land of fish and rice" with numerous lakes and networks of rivers and canals.

The river basins in China are customarily divided into outflow and inflow drainage areas. The outflow area covers 6,121,300 sq.km. and the inflow area 3,478,700 sq.km., accounting for 63.8% and 36.2% of the total area of the country respectively. The Changjiang (Yangtze) River and the Huanghe (Yellow) River are the two largest rivers in China. There are also a large number of lakes with a total area over 80,000 sq.km. The five largest freshwater lakes of China (APH 1983) are the Boyang Lake, the Dongting Lake, the Taihu Lake, the Hongze Lake and the Chaohu Lake.

1.1.2 Climate

Among climatic factors of central importance in China are the pattern of precipitation and the natural temperature range, because together they determine the suitability of arable land for different patterns of cultivation.

Rainfall The average annual rainfall in China is 630 mm. China's climate is broadly monsoonal, which means that approximately 80% of the rainfall comes in the four warmest months. The advantage of such a pattern for agricultural production is the coincidence of the rainfall with the height of the growing season which has the greatest water demands. However, the monsoonal climate which is extremely variable, both in total amount and in the seasonal pattern, creates obvious difficulties for farming. The rains may come too late or too early or are too widely spaced. Under these circumstances, soil may dry out and crops die or are stunted. Rain may also be violent and excessive, which can cause destructive floods and also leave a large part of the growing season overly dry. Variability is the greatest where average rainfall is the least. Generally, the further from the sea, which is the origin of the moisture bearing airmasses of the summer monsoon, the less the average annual rainfall and the greater the variability (Murphey 1982, pp54,56).

Annual precipitation in the south eastern provinces is about 1,600 mm, 50–60% of which falls in the monsoon season in normal years; whereas provinces of northwestern China have less than 200 mm precipitation in most years, 70–80% of which falls in the monsoon season.

Temperature However, it is notable that for the same latitude and altitude, the number of frost-free days are fewer and thus the growing season shorter in China than elsewhere in the Eurasian Continent (Shen 1951, Ch 2; Leeming 1985, Ch 2). Six climatic zones have been defined by using the continuous period of daily mean temperature 10°C as the criterion (APH 1983, p23) these are summarized in Table 1.1.

1.1.3 Water resources

China is rich in water resources. Estimates show that the country's annual precipitation contributes about 6,000 billion cubic metres of water, which makes up 5%

Table 1.1 Climatic zones of China

Zone	Location	Period over 10°C†	Cumulative temperature	Annual precipitation	Continuous frost-free period	Cropping pattern
1 The north temperate zone	The north of Heilongiang Province	<100 days	<1600°C	350–500mm	3 months	One crop per year
2 The middle temperate zone	Most of N-E China, Nei Mongal and N Xingiang	100–160 days	1600–3400°C	400–1000mm (Coastal) 100–400mm (Inland)	4–7 months	One crop per year
3 The south temperate zone	North China & S Xingiang	100–160 days	3400–4500°C	400–800mm (N China) <100mm (S Xiangiang)	5–8 months	Two crops/ yr or three crops/ 2 years
4 The sub-tropical zone	Region S of Qining Mt and N of Naning Mt.	210–365 days	4500–8000°C	1000–2000mm	8–12 months	Two crops/ yr sub-tropical industrial crops
5 The Tomia zone	Leizhou Peninsula, S China Sea Islands	350–365 days	≥8000°C	1400–2000mm	Whole year	Tropical crops
6 Plateau region	Qinghai-Xizang Plateau		<2000°C			Crops only below 4000m

Source: Derived from p23 of Agriculture of China, Agricultural Publishing House ed. (1983) Beijing China

†Continuous period with daily temperature >10°C

of the total global precipitation, and the run-off water from streams and rivers is about 2,600 billion cubic metres, which accounts for 5.5% of the total global run-off water (APH 1983, p27). Although the gross amount of water may seem adequate, its spatial and temporal distribution handicap the healthy development of Chinese agriculture.

The spatial distribution of river water and the farmland available for cultivation do not match well. Generally, there is more water in the southeast and less in the northwest, so there is a gradual decrease in water reserves from southeast to northwest. For example, while the north and the northwest have 51% of the total arable land, they receive only 7% of the total river discharge (Greer 1979, p55).

The uneven temporal distribution of water resources is another major problem. The three major rivers in north China, the Huaihe, the Huanghe (Yellow) and the Haihe rivers have highly variable flows not only from month to month but also year to year. As a consequence, floods and droughts have been the main natural disasters for agriculture since time immemorial (Hussain 1986). This explains the priority given to water conservancy in improving agricultural production conditions since the founding of the People's Republic of China (hereafter PRC) (see section 1.2).

1.1.4 Land resources and utilization

Combining the physical features of geology, topography, climate and soil with the distribution of population defines three characteristics of Chinese agricultural land. These are: the scarcity of arable land, its uneven quality and the uneven distribution of good land.

Land scarcity Given China's large population, its land resources are extremely low per head. This is best indicated by comparing the areas per capita of different types of land in China with the corresponding world averages (Table 1.2).

Furthermore, of the 500 million mu (over 33 million hectares) of reclaimable land in the country, only 150 million mu (10 million hectares) are of relatively good quality. Therefore, there is little room for land reclamation.

Land quality Of the existing 1,500 million mu (100 million hectares) of total cultivated land in China, only two thirds is good or fair. The area suitable for

Table 1.2 Land area available in China and the World

	China	World average
	ha/head	
Total land area	< 1	3.3
Cultivated land	0.1	0.34
Forest area	0.13	1.03
Grassland	0.34	0.76

Source: derived from CHEC 1984, p206.

irrigation is 720 million mu (48 million hectares), less than half of the total. There are 100 million mu (about 6.7 million hectares) of saline and alkaline land, 140 million mu (over 9.3 million hectares) of dry land is affected by wind and sandstorm, 100 million mu (about 6.7 million hectares) suffers from serious soil erosion, and 180 million mu (12 million hectares) is low-yielding red soil (China Handbook Editorial Committee 1984 (hereafter CHEC 1984), p204).

Land distribution The distribution and utilization of land resources in China reveal sharp regional differences. Over 90% of the cultivated land, forest and water area lie in the southeastern humid and semi-humid part of the country. This is the main region where crop farming, forestry, sideline production and fisheries are concentrated. The vast grasslands and more difficult areas are concentrated in the arid and semi-arid northwestern region where animal husbandry is the main agricultural activity. The current land utilization pattern in China is shown in Table 1.3.

1.2 The Improvements of Production Conditions

Despite the long history of agricultural civilization in China, the country does not possess favourable natural conditions for farming. Since the founding of the PRC, capital accumulation in agriculture has been encouraged on a massive scale for the improvement of production conditions. These efforts were concentrated on water conservation, mechanization and the use of chemical fertilizer (CHEC 1984).

Water conservation For thousands of years Chinese agricultural production was severely affected by droughts and floods. Historical records show that in the 2,000 years from 206 B.C. to 1949, the country was hit by 1,092 major floods and 1,056 major droughts i.e. every other year on average there is a major flood and a major drought. Over the period 1949 to 1980, 47.3 billion yuan (US$13.24 billion) was allocated by the Chinese government for the construction of large scale water conservancy projects. At the same time, about 58 billion yuan was contributed to

Table 1.3 A summary of land utilization

Proportions of total land area which is:	
1. Cultivated land	10.4%
2. Fruit, tea, mulberry and rubber plantations	0.3%
3. Forestry	12.7%
4. Grassland	33.0%
5. Urban and industrial areas and transport routes	6.9%
6. Inland waterways and shallow seas	3.1%
7. Stone bareground, deserts, marshes, permanent frozen land and glaciers	19.4%

Source: from CHEC 1984, p203.

this investment by the collective production units by raising funds and providing labour. During this period, flood-prevention projects were built on the Huanghe River, Huaihe River, Haihe River and Changjiang River. This included more than 86,000 reservoirs of varying sizes, over 6.4 million ponds with a total storage capacity of 400 billion cubic metres, and more than 5,000 irrigation projects each benefiting, on average, 650 hectares of farmland. In 1980, the irrigated area totalled 44.6 million hectares, accounting for nearly half of the total arable land. This gives China one of the highest irrigation ratios in the world.

Mechanization Although agricultural activities in China are traditionally highly labour intensive a steady process of mechanization has been underway by increasing the supply of electricity and petroleum to agriculture. By the end of 1980, the stock of mechanized farm power totalled 198 million hp including 745,000 tractors, 1,874,000 walking tractors and 27,000 combine harvesters. The machine-ploughed areas totalled 42 million hectares, accounting for 41.3% of all farmland, and the machine-sown area accounted for 10.9% of the total. The increase in the level of mechanization has been one of the critical factors enabling the sustained increase in intensity of farming shown by a rise in the cropping index over the last three decades.

The use of chemical fertilizer Increasing use of chemical fertilizer has been the other main contributor to the rapid rise of grain crop yields worldwide since the Second World War. In China, chemical fertilizer has had its greatest impact on yields since the 1960s. It is expected that further increases in the use of fertilizers will be one of the main ways of achieving continued increases in farming output in the future (Barnett 1981, p331).

The chemical fertilizer industry was set up in 1949 and has grown rapidly since then. During the Fourth Five-Year Plan period (1971-75), 13 large scale chemical fertilizer plants were imported by the Chinese government. These big enterprises, together with more than 1,300 small nitrogenous fertilizer plants and around 50 medium-sized ones, form a network of nitrogenous fertilizer production throughout the country. Chemical fertilizer consumption in 1952 amounted to just 78,000 tonnes, by 1983, 16.6 million tonnes was applied.

These changes in agricultural production conditions are summarized in Table 1.4.

1.3 The Historical Development of Chinese Agricultural Production

1.3.1 Agricultural production

In 1952, the gross current value of agricultural output was estimated as 46.3 billion yuan, which made up 45% of the GNP of the country. In 1984, the output value of agriculture reached an unprecedented level of 375.5 billion yuan, over eight times

Table 1.4 Changes in agricultural production conditions

	1952	1983
Total reservoir capacity (million cubic metres)	30,000	420,000
Capacity of irrigation & drainage equipment (million hp)	1.2	786.6
Area under irrigation (million hectares)	19.96	46.64
Large and medium-sized tractors	13,000	841,000
Heavy trucks for agricultural use	280	274,800
Total power capacity of farm machinery (million hp)	0.25	245
Electricity consumed (million kw)	50	43,520
Fertilizer applied (1,000 tons)	78	16,598
Tractor-ploughed farmland (million hectares)	1.4	33.57

Source: Du Runsheng, *China's countryside under reform.* In Beijing Review, Vol. 27, No. 33, p17, August 13, 1984.

as much as that of 1952, with an annual growth rate of 6.8%. However this output accounted for 29% of the GNP in that year (calculated in current prices, from the State Statistical Bureau ed. 1985 (hereafter SSB 1985) pp20,23).

Taking 1952 as the base year, the index of real gross value of agricultural output increased to 394, with an annual growth rate of 4.4% between 1952 and 1984 (SSB 1985, p25). The trends and structure of agricultural and industrial development are shown in Chart 1.1. This indicates the more rapid growth in industrial output and thus the declining share of agricultural output in the total.

China's agricultural production is traditionally classified into five areas: crop farming, forestry, animal husbandry, fisheries and sideline occupations. Sideline occupations include collecting wild plants, hunting wild animals, peasant households' handicrafts and rural industry run by basic production units (production brigades and production teams). The make-up of the five components is shown in Table 1.5 and Chart 1.2. These show that crop farming has remained the biggest portion of Chinese agricultural production since the 1950s although its proportion has decreased over time. The growth of output for the major agricultural products and the trend of growth is shown in Table 1.6 and Chart 1.3 respectively. The most striking features of these trends is the acceleration in growth of major industrial crops and meats since 1978. An unprecedented growth in grain production was also achieved during the same period. The policy changes which precipitated these developments will be reviewed in detail in later chapters.

1.3.2 Grain production

Grain crops in China include: rice, wheat, barley, oats, rye, corn, sorghum and millet. In China it is common usage to categorize tubers (sweet potato, potato) and legumes (soybean, broad bean, pea, green gram, red bean) as grain crops. Among these crops rice, wheat, corn, soyabeans and sweet potatoes are the most important. As grain crops have always taken up around 80–90% of the nation's total farmland,

CHART 1.1 Indices of the Value of Gross Output by Industrial Sector, 1952–1984

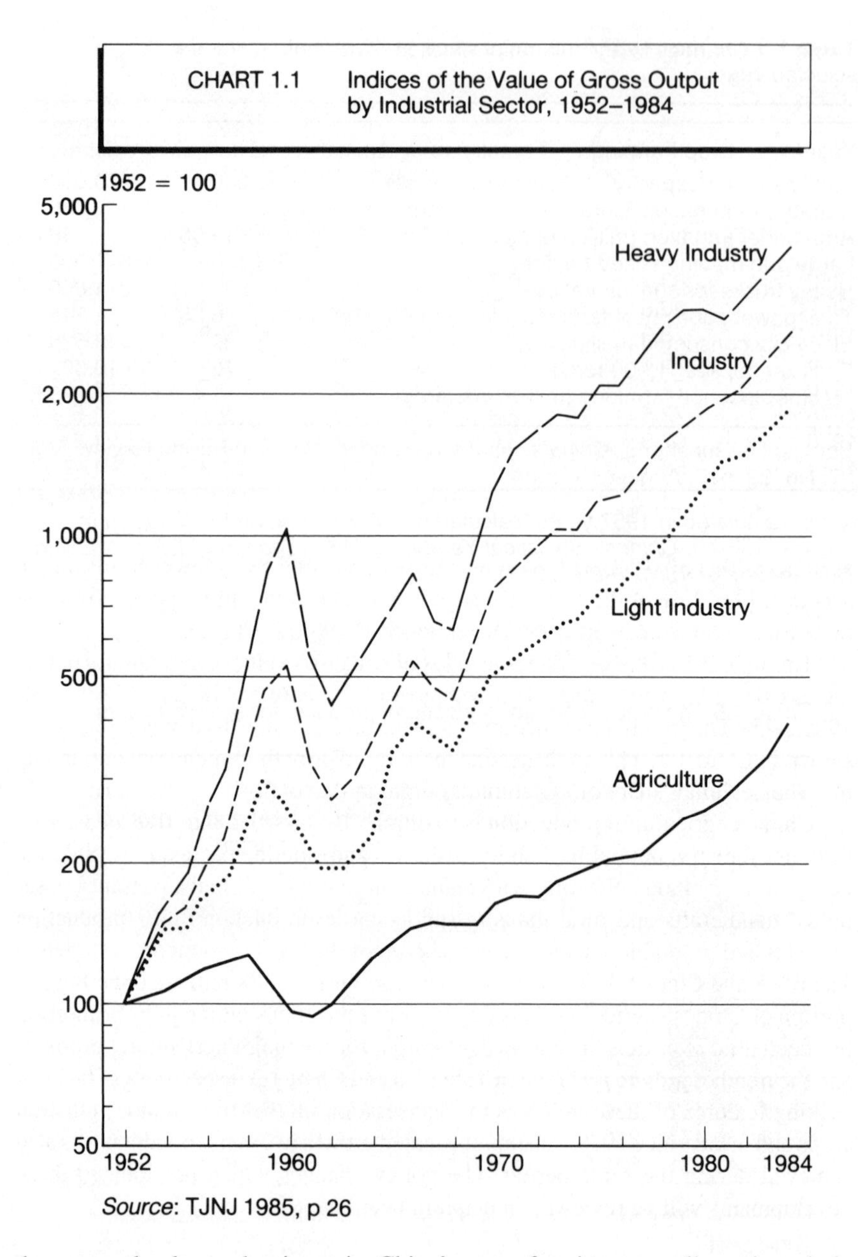

Source: TJNJ 1985, p 26

these are clearly predominant in China's crop farming as well as the whole agriculture sector.

1.3.2.1 The change in growing area

Two basic features characterize the historical changes in grain growing areas; these are shown in Table 1.7. First, the total growing area for grain crops has been declining as indicated both by the real cultivating area and the proportion of grain

Table 1.5 The make-up of the gross value of China's agricultural output for selected years

Year	Crop Farming	Forestry	Animal Husbandry	Sidelines	Fishery
			Percentage of output		
1952[a]	83.1	0.7	11.5	4.4	0.3
1957[a]	80.6	1.7	12.9	4.3	0.5
1962[a]	78.9	1.7	10.3	7.3	1.8
1965[a]	75.8	2.0	14.0	6.5	1.7
1970[a]	74.7	2.2	12.9	8.7	1.5
1975[b]	72.5	2.9	14.0	9.1	1.5
1978[b]	67.8	3.0	13.2	14.6	1.4
1984[c]	58.1	4.1	14.2	21.9	1.7

Note: [a]calculated in 1957 yuan, [b]calculated in 1970 yuan, and [c]in 1980 yuan.
Source: SSB ed. *Chinese Statistical Yearbook, 1985 (Zhongguo Tongji Nianjian 1985* (hereafter *TJNJ 1985*), p241).

Table 1.6 The output of major agricultural products for selected years

Year	Grain (1)	Cotton (2)	Oil-bearing Crops (3)	Sugar-yielding Crops (4)	Pork, beef and mutton (5)	Aquatic products (6)
			(*in 10,000 tonnes*)			
1952	16,392	130.4	419.3	759.6	338.5	167
1957	19,505	164.4	419.6	1,189.3	398.5	312
1962	16,000	75.0	200.3	378.3	194.0	228
1965	19,453	209.8	362.5	1,537.5	551.0	298
1970	23,996	227.7	377.2	1,556.0	596.5	318
1975	28,452	238.1	452.1	1,914.3	797.0	441
1978	30,477	216.7	521.8	2,381.8	856.3	466
1984	40,731	625.8	1,191.0	4,780.3	1,540.6	619

Note: the output for sugar-yielding crops is the sum of sugarcane and sugarbeet.
Source: Column (1), (2) and (3) - *TJNJ 1985*, p257
Column (4) - *TJNJ 1985*, p258
Column (5) - *TJNJ 1985*, p267
Column (6) - *TJNJ 1985*, p270

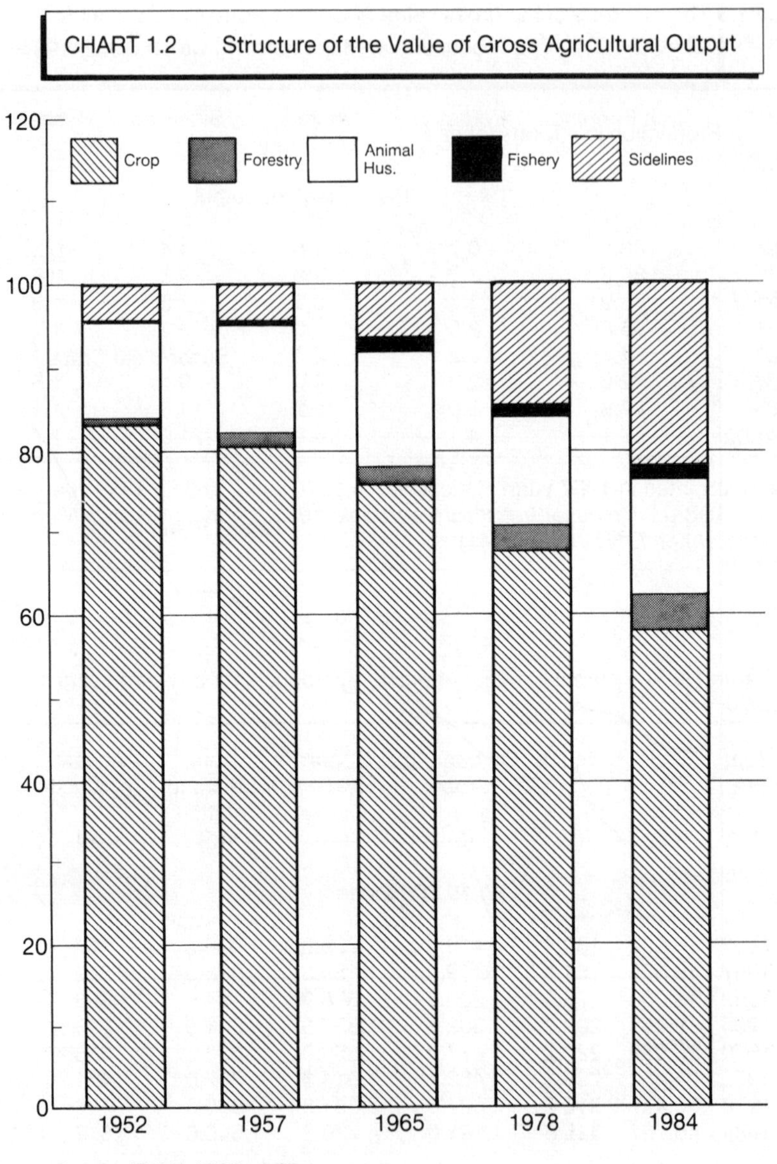

CHART 1.2 Structure of the Value of Gross Agricultural Output

Source: TJNJ 1985, p 26

growing area in the total farming area. From 1952 to 1984, the grain growing area decreased from 1,859.7 million mu (123 million hectares) to 1,693.3 million mu (113 million hectares), a net reduction of 10 million hectares. At the same time the proportion of the total growing area devoted to grain decreased 9.5 % from 87.8% to 78.3%. This change has been more pronounced since the implementation of the new development policy in 1978. About three quarters of the reduction in grain growing area from 1952 to 1984 took place between 1978 and 1984.

CHART 1.3 The Production of Major Farming Crops, 1952–1984

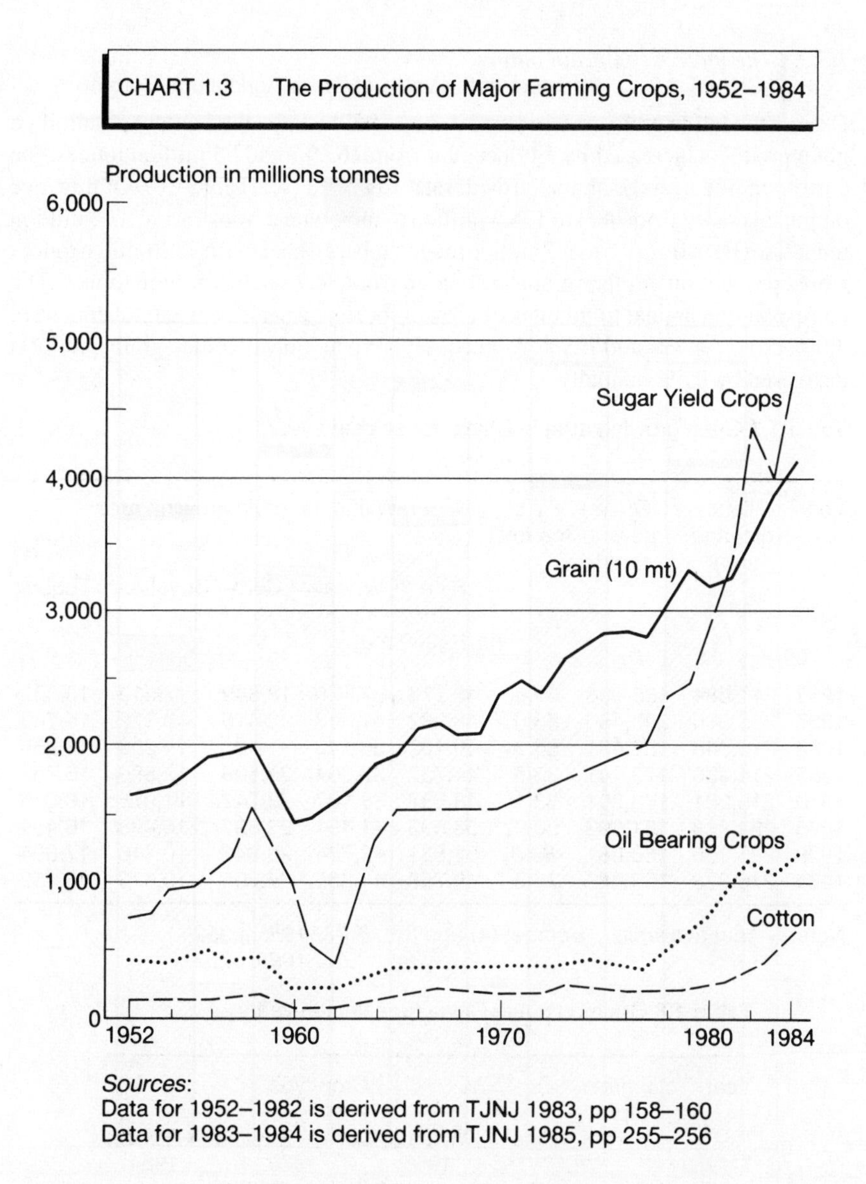

Sources:
Data for 1952–1982 is derived from TJNJ 1983, pp 158–160
Data for 1983–1984 is derived from TJNJ 1985, pp 255–256

Second, while the total grain growing area was decreasing, the growing areas for high yielding crops increased. For example, the rice growing area increased from 425.7 million mu (28.4 million hectares) to 497 million mu (nearly 33.2 million hectares) the growing area for wheat rose from 371.7 million mu (around 24.8 million hectares) to 443.7 million mu (about 29.6 million hectares), and the corn growing area increased from 188.5 million mu (nearly 12.6 million hectares) to 278.1 million mu (about 18.5 million hectares). The net increase in growing area for these three crops is thus 4.8, 4.8 and 5.9 million hectares respectively, totalling 15.5 million hectares.

1.3.2.2 The increase in grain output

Despite the fall in grain growing area during 1952–84, the total output of unmilled grain products increased by 149 per cent from 163.9 to 407.3 million tonnes.The corresponding average annual growth rate was 2.9% (see Table 1.8). Of this, rice output increased from 68.4 to 178.3 million tonnes, wheat from 18.1 to 87.8 million tonnes, soybean from 9.5 to 9.7 million tonnes, tubers from 16.3 to 28.5 million tonnes whilst production of other grains decreased from 34.7 to 29.6 million tonnes. The corresponding annual trend rates of growth for rice, wheat, corn and tubers were 3.0%, 5.1%, 4.7% and 1.7% respectively. Production of other grain products decreased by 0.5% annually.

Table 1.7 Grain growing area in China, for selected years

Year	Total growing area[a]	Grain growing area[b]	% of the total	Within the grain growing area[c]				
				Rice	Wheat	Corn	Soyabean	Tubers
				(in 10,000 mu)				
1952	211,884	185,968	87.8	42,573	37,170	18,849	17,519	13,032
1957	235,866	200,450	85.0	48,362	41,313	22,415	19,122	15,742
1962	210,343	182,431	86.7	40,402	36,113	—	14,256	18,256
1965	214,936	179,441	83.5	44,737	37,064	23,506	12,889	16,763
1970	215,231	178,901	83.1	48,537	38,187	23,747	11,978	16,076
1975	224,318	181,593	80.9	53,593	41,491	27,897	10,498	16,454
1978	225,156	180,881	80.3	51,631	43,774	29,942	10,716	17,694
1984	216,332	169,326	78.3	49,768	44,365	27,805	10,929	13,482

Note "—" not available Source: (a) and (b) : *TJNJ 1985*, p252
(c) : *TJNJ 1985*, p253.

Table 1.8 Grain output in China for selected years

Year	Total grain output	Crop type				
		Rice	Wheat	Corn	Soyabean	Tubers
				(in 10,000 tonnes)		
1952	16,392	6,843	1,813	1,685	952	1,633
1957	19,505	8,678	2,364	2,144	1,005	2,192
1962	16,000	6,299	1,667	—	651	2,345
1965	19,453	8,772	2,522	2,366	614	1,986
1970	23,996	10,999	2,919	3,303	871	2,668
1975	28,452	12,556	4,531	4,722	724	2,857
1978	30,477	13,693	5,384	5,595	757	3,174
1984	40,731	17,826	8,782	7,341	970	2,848

Source: *TJNJ 1985*, p255.

Two main features emerge from this summary of the historical development of Chinese grain production. First, is the increase in grain output, which occurred mostly through the growth in yield. Much of the yield improvement was due to the shift of growing area from low-yielding to high-yielding crops as discussed above. As China has such a long history of agricultural civilization and a traditionally high population/land ratio, increasing output by means of bringing more land into cultivation has not been an available option for a very long time. Therefore, the drive for output growth has been focussed on increasing the intensity of agriculture aimed at achieving higher yield. Since the 1950s, the improvement of grain yield has been achieved by means of water control (improved irrigation and drainage), increased use of chemical fertilizers, more responsive and thus higher yielding seed varieties and mechanization (see Timmer 1976; Barnett 1981; World Bank 1985b). This large injection of modern inputs is the main distinction from the traditional intensive farming which existed in China for centuries. The yield improvements for major grain crops in the last three decades are summarized in Table 1.9.

The second feature which can be seen in Charts 1.3 and 1.4, is that the historical development of China's agricultural production has followed an erratic course since the founding of the PRC. As there have been no dramatic shocks in inputs use or technological change, the shifts in government policies are suspected as a principal factor explaining the observed development pattern. These policy issues will be reviewed in the next chapter and, indeed, they form the main subject of this book.

1.4 Food Consumption

During the last three decades, China's experience with food consumption has been similar to that of other developing countries in most respects. The key determining factor for the growth in grain consumption "is the combined result of rapid popu-

Table 1.9 Yields for major grain crops 1952-84

Year	Grain crops as a whole	Crop type				
		Rice	Wheat	Corn	Soyabean	Tubers
		(tonnes/ha)				
1952	1.32	2.41	0.73	1.34	0.82	1.88
1988	3.61	5.37	2.97	3.96	1.33	3.17
Total increase (times)	2.73	2.23	4.07	2.96	1.62	1.69
Average annual growth rate (%)	3.2	2.5	4.5	3.4	1.5	1.7

Source: derived from Tables 1.7 and 1.8.

CHART 1.4 The Trend of Grain Production, 1952–1984

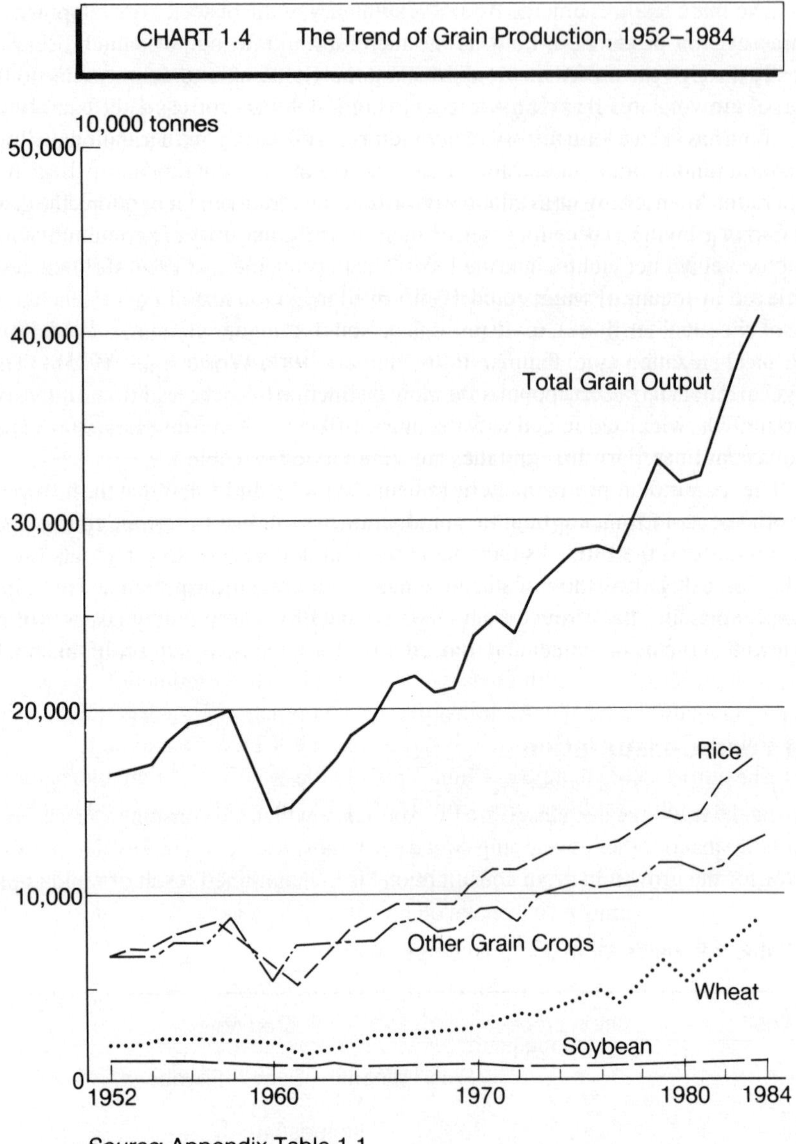

Source: Appendix Table 1.1

lation increase and a high income elasticity of demand for grain at prevailing low levels of income per head" (Walker 1984). Each of these will be reviewed in turn.

1.4.1 Population growth

China is the most populous nation in the world. According to the census carried out on July 1st, 1983, the total population of the PRC was 1,008.2 million persons, which accounted for more than one fifth of the total world population.

The three main characteristics of China's population since the 1950s are the speed of growth, the transitions it has undergone and the age structure.

Table 1.10 shows the increase in total population in China from 574.8 million persons in 1952 to 1,034.8 million persons in 1984. This corresponds to an annual natural growth rate of nearly 1.9%, which is significantly higher than developed countries and lower than that found in many developing countries in Africa, Asia and Latin America (Wang 1985, p805). However, given the large population base in China a 1% annual natural growth rate means a net increase of more than 10 million people per annum. During the 32 years from 1952 to 1984, the annual net increase in population registered 10–15 million persons for 15 years (1953–56, 1958–59,1962, 1975–82), 15–20 million persons for seven years (1957, 1963,1965–67, 1972–73), and more than 20 million persons for four years (1968–71) (*TJNJ 1985*, p185). The highest population growth took place during the 1950s, 1960s and early 1970s, which coincided with the two well-known socio-political phenomena, the "Great Leap Forward" and the "Cultural Revolution".

The population growth pattern in China typifies the phenomenon of the first transition period. According to population growth theory, population growth processes for any society usually pass through three stages, from high birth, high death rate to high birth, low death rate and then, finally to the low birth and low death rate (Sung et al., 1982). Since the founding of the PRC the continuing improvement in health services has enabled a transition from a pattern of high birth, high death rate to high birth, low death rates (see Table 1.11). The combination of the huge initial population base and this transition in the population growth pattern has thus been the major drive for the accelerated increase in China's population.

The third striking feature of China's population issue is the premature nature of the age structure. As can be seen from Table 1.12, 46.1% of the total population in

Table 1.10 Population growth in China for selected years

Year	Total population	Population location	
		Cities & towns	Countryside
	(Population in millions)		
1952	574.82	71.63	503.19
1957	646.53	99.49	547.04
1962	672.95	116.59	556.36
1965	725.38	130.45	594.93
1970	829.92	144.24	685.68
1975	924.20	160.30	736.90
1978	962.59	172.45	790.14
1984	1,034.75	330.06	704.69

Source: *TJNJ 1985*, p185

Table 1.11 The birth, death and natural increase rate

Year	Birth rate	Death rate	Natural increase rate
(rates per thousand)			
1952	37.00	17.00	20.00
1957	34.03	10.80	23.23
1962	37.01	10.02	26.99
1965	37.88	9.50	28.38
1970	33.43	7.60	25.83
1975	23.01	7.32	15.69
1978	18.25	6.25	12.00
1984	17.50	6.69	10.81

Source: *TJNJ 1985*, p186

Table 1.12 The age structure in China, July 1st, 1983

Age groups	Total population (persons)	Percentage of the total (%)
0 - 19	462,617,533	46.1
20 - 39	293,106,748	29.2
40 - 59	170,551,102	17.0
60 -	76,638,524	7.6
Total	1,003,913,927	100.0

Source: *TJNJ 1985*, p198

mid 1983 were under the age of 19, and more than three quarters of the population were less than 40 years old. This factor ensures a continuing high birth rate and thus national population increase for the next few decades.

1.4.2 Income growth

Poverty and hunger have loomed over the heads of Chinese people for a very long time, right up to the present. China was a semi-colonial and semi-feudal society before the founding of the PRC. The destruction by a succession of wars during the first half of the century had left the country backward. Per capita disposable income was thus very low at the beginning of the 1950s. Indeed for most of the period until the end of the 1970s, disposable income growth was constrained by the policy of maximizing capital accumulation in order to speed the pace of industrialization. It is only in the very short period since 1978 that a new trend has emerged. This is associated with the commitment by the Chinese government to improve the income and daily living standards of the population.

The data summarized in Table 1.13 show that for the 26 years from 1952 to 1978, the per capita net cash income of the nation increased from 47.4 yuan to 131 yuan (in current prices), an average increase of 4% per annum. However, in the five years from 1978 to 1983 per capita net cash income rose from 131 yuan to 259 yuan, an annual growth rate of 14.6%.

Since the net cash income is only part of the net real income for Chinese peasant households, the per capita net real income of the peasant households from a sample survey is presented in Table 1.14.

1.4.3 Food consumption

The general picture of food consumption in China since the founding of the PRC is shown in Table 1.15. The basic features of food consumption for the last three decades can be summarized as, a low average consumption level with large temporal fluctuations; large disparities inter-regionally and between cities, towns and the countryside; and low levels of energetic and nutritional contents from animal sources.

Table 1.13 Net cash income per head in China

	Location					
	Average of the nation		Cities and towns		Countryside	
Year	In current prices	In 1952 constant price	In current prices	In 1952 constant price	In current prices	In 1952 constant price
	(yuan per annum)					
1952	47.4	47.4	183.2	183.2	30.0	30.0
1957	69.3	63.9	242.2	223.0	39.7	36.9
1962	79.6	58.3	296.4	203.0	40.3	31.6
1965	84.8	70.4	291.9	236.7	46.2	39.1
1970	86.7	73.7	314.6	255.2	48.9	43.0
1975	110.1	93.3	401.5	309.3	60.1	54.6
1978	131.0	107.7	440.8	312.0	70.4	63.9
1983	258.7	185.7	637.5	383.3	175.1	142.0

Note: Figures under 1952 constant price columns are the net cash incomes in current prices deflated with retail price indices (1952 = 100) for the nation, city and town and countryside respectively.
Source: From *China's Statistical data on Trade and Prices, 1952-1983 (Zhongguo Caimao Wujia Tongji Ziliao, 1952-1983*, hereafter *ZCWTZ 1952-1983)* pp20, 370.

Table 1.14 Per capita net income of peasant households

Per capita net income			Per capita net income		
Year	In current prices	In 1952 constant price	Year	In current prices	In 1952 constant price
	(in yuan)			(in yuan)	
1957	72.95	67.7	1981	223.44	186.5
1978	133.57	121.2	1982	270.11	221.8
1979	160.17	142.5	1983	309.77	251.2
1980	191.33	163.0	1984	355.33	—

Note: Figures under 1952 constant price are the net incomes in current prices deflated with countryside retail price index (1952=100).
Sources: Data for per capita net income is from *TJNJ 1983*, p499, and *TJNJ 1985*, p571
Data for the retail price index is from *ZCWTZ 1952-1983*, p370, p32.

In 1983 the average Chinese consumed (directly) about 232 kg of cereals, 4 kg of vegetable oil, 24.3 kg of pork meat (which accounts for about 95% of the total meat consumption in China), 9 kg of sugar and 3 kg of eggs. Grain therefore makes up around 90% of the nutritional intake in China. Despite this heavy reliance on grain and despite the official policy for most of the period prior to 1978 which gave priority to grain production - the so-called, "take grain as the link" policy - the per capita grain consumption level decreased from 198 kg (1952) to 195 kg (1978). For most of the time during the 1950s to the 1970s, per capita grain consumption was below 200 kg with the lowest level recorded in 1961 when consumption was below 160 kg (SSB 1984, p27). Although the total output of unmilled grain products increased from 163.9 million tonnes in 1952 to 387.3 million tonnes in 1983, an annual growth rate of 2.8%, the high population growth of 1.9% per annum over the same period meant that the average per capita grain consumption increased by just 34 kg from 198 kg to 232 kg. The annual rate of improvement was just 0.5%. The temporal fluctuation of per capita grain consumption is shown as an index in Chart 1.5.

Although the Chinese food distribution system has been successful in providing basic food nutrients to most of the people most of the time, inter-regional differences in per capita food consumption and nutrient availability are still apparent. According to the World Bank (1985b, p17), per capita grain production ranged from 202kg (Guizhou) to 420 kg (Jiangsu) in 1981; average per capita availability of protein ranged from about 38 g (Guizhou) to more than 100 g (Heilongjiang) in 1980; and average per capita meat production in 1980-81 ranged from 19.3 kg in Sichuan to 5.4 kg in Ningxia.

Table 1.15 Per capita consumption of major food products, for selected years

	Milled grain (a)			Vegetable oil (b)		
Year	National average	City & town	Country-side	National average	City & town	Country-side
		(Jin)			(Jin)	
1952	395.34	480.70	383.44	4.19	10.23	3.35
1957	406.12	392.00	408.76	4.84	10.30	3.71
1962	329.25	367.67	321.13	2.17	4.92	1.59
1965	365.67	421.30	354.25	3.43	9.63	2.17
1970	374.43	403.58	368.75	3.21	8.47	2.19
1975	381.03	418.51	373.87	3.45	9.31	2.33
1978	390.92	410.58	386.66	3.19	8.21	2.10
1983	464.46	443.35	469.41	8.05	19.95	5.25
	Pork (c)			Sugar (d)		
Year	National average	City & town	Country-side	National average	City & town	Country-side
		(Jin)			(Jin)	
1952	11.83	17.84	10.99	1.82	5.99	1.24
1957	10.15	17.95	8.70	3.02	7.26	2.22
1962	4.43	7.58	3.77	3.19	7.04	2.38
1965	12.57	20.73	10.89	3.36	7.08	2.59
1970	12.03	21.50	10.19	4.11	8.68	3.22
1975	15.25	29.84	12.46	4.52	11.31	3.22
1978	15.34	27.39	12.73	6.84	16.29	4.79
1983	24.69	36.08	22.02	8.94	19.07	6.56
	Fresh eggs (e)					
Year	National average	City & town	Country-side			
		(Jin)				
1952	2.04	3.51	1.84			
1957	2.51	3.99	2.24			
1962	1.53	1.42	1.56			
1965	2.84	4.12	2.58			
1970	2.64	3.36	2.50			
1975	3.26	2.96	3.32			
1978	3.94	3.93	3.94			
1983	5.92	7.34	5.59			

Note: 2 Jin = 1 kg.

Sources: Column (a): *ZCWTZ 1952-1983*, p27; Column (b): *ZCWTZ 1952-1983*, p28; Column (c): *ZCWTZ 1952-1983*, p29; Column (d): *ZCWTZ 1952-1983*, p30; Column (e): *ZCWTZ 1952-1983*, p31

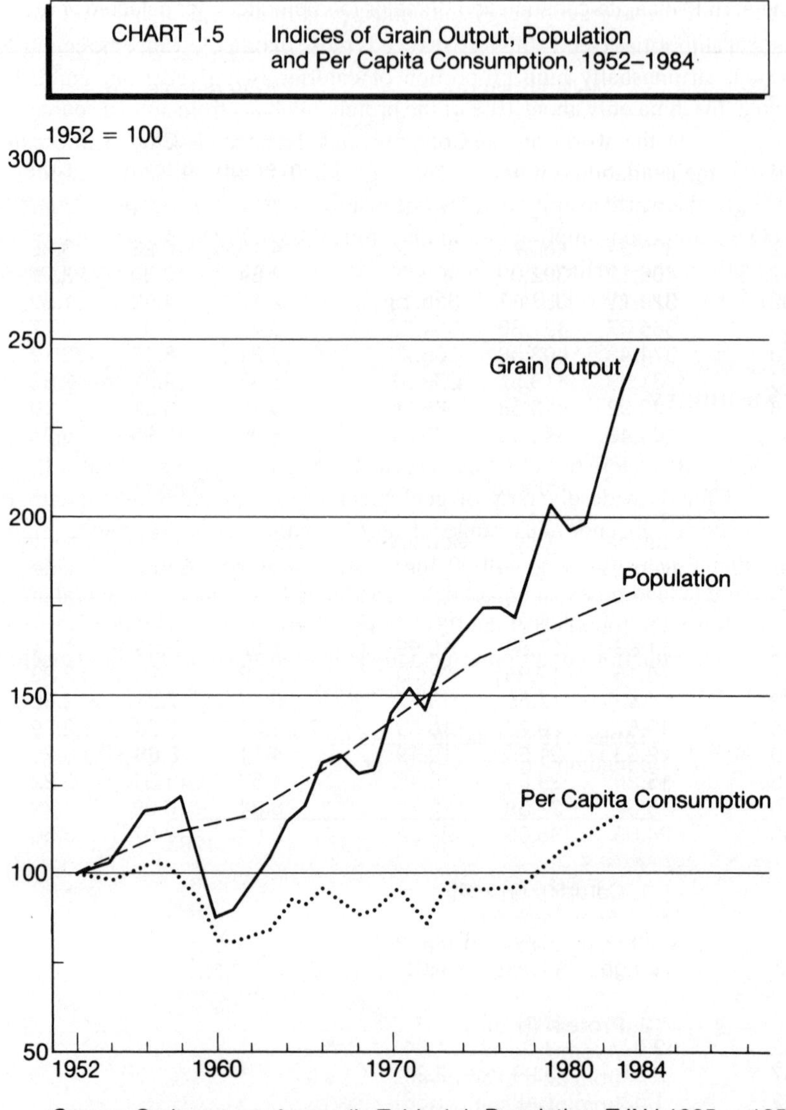

CHART 1.5 Indices of Grain Output, Population and Per Capita Consumption, 1952–1984

Source: Grain output, Appendix Table 1.1, Population, TJNJ 1985, p 185.
Per capita grain consumption, Appendix Table 1.6

Another dimension of the disparity in per capita food consumption is between city and town and countryside. Take 1978 as an example, although the population in the countryside accounted for 82.1% of the total population (Table 1.10) and it is this population who supply the food, their food consumption level is much lower than those in city and town. The per capita consumption in the countryside is 193.3 kg for grain, 1.05 kg for vegetable oil, 6.4 kg for pork meat and 2.4 kg for sugar, which represents 94%, 26%, 46% and 30% of the consumption level in the city and town areas respectively (Table 1.15).

Although the average energy and nutritional intake calculated by Chinese statistical authorities (see Table 1.16) exceeds those of other low income countries, in China, an unusually high proportion of calories is derived from grain. For example, in China only about 10% of the protein intake is from animal sources, as compared with the world average of 35% and 21% for developing countries. Similarly, the availability of animal and vegetable fat in 1980–82 was more than 40% below the world average and 5% below the average of developing countries. Direct per capita consumption of grain in China (about 209 kg p.a. in 1980–82) is amongst the highest in the world and exceeds that in India by 60% and in Indonesia by more than 30% (World Bank 1985b, pp17–18).

1.5 Summary

In short, the characteristics of China's agriculture can be summarized as follows:

(1) China's wide diversity of geological, topographical, climatic and soil features, create an enormous range of agro-climatic zones and thus a highly differentiated agriculture. In addition, the huge population in relation to the slender base of arable land poses an unparalleled challenge to Chinese agriculturalists.

(2) Since the founding of the PRC, huge efforts have been made to improve agricultural production conditions especially irrigation and drainage, farm machin-

Table 1.16 Nutritional intake from food consumption, per capita/per day

	1978	1983
1. Calories (in kcal)		
Average level	2,311.0	2,877.4
from animal sources	142.0	225.9
from plant sources	2,169.0	2,651.5
2. Protein (in g)		
Average level	70.8	82.6
from animal sources	4.0	6.2
from plant sources	66.8	76.6
3. Fat (in g)		
Average level	29.9	47.2
from animal sources	13.9	22.1
from plant sources	16.0	25.1

Source: *TJNJ 1985*, p58.
Note: It is generally accepted that these data based on foodstuff availability for consumption tend to overstate actual intake due to losses in distribution and storage. This is confirmed by nutritional surveys.

ery and the use of chemical fertilizers. This involved massive capital investment which has been the main contributor to the substantial increases in the output of grain and other agricultural products during the last three decades.

(3) Crop farming has always been the biggest sector of China's agriculture since the 1950s, although its proportion decreased with time. Two main features of grain production were identified. First was the substantial growth in grain output achieved between 1952 and 1984. This is partly due to the continuing shift of growing area from low-yielding to high-yielding crops and, more importantly, induced by the large injection of modern inputs in production with the consequences of a significant increase in crop yields. The second feature was that the production of grain and numerous other crops followed a zig-zag course since the founding of the PRC. The main explanation for this development pattern is thought to be shifts in government policy. This is the central focus of subsequent chapters.

(4) As in other countries, food consumption in China is fundamentally determined by the status of population growth and the improvement in per capita disposable income. Due to the high population growth on the one hand and only a marginal improvement in per capita disposable income on the other, the basic features of the nation's food consumption pattern can be summarized as having a low average consumption level with large temporal fluctuations; large disparities inter-regionally and between city and town and countryside; and low levels of energetic and nutritional contents derived from animal sources.

1.6 General Assessment of the Statistical Data Concerning China's Agriculture

Studies on the development of Chinese agriculture have always been troubled by data problems. There are three broad sets of issues involved. The first concerns data availability. For well-known reasons, official statistics of China's development were not made available to foreign researchers for about two decades from the late 1950s. Consequently figures collected and pieced together by various foreign and international organizations were usually used. Those from the United States government agencies mainly involve the Central Intelligence Agency (CIA), the Department of Agriculture (USDA) and the Department of Commerce (DOC) (see Tang and Stone 1980). The principal international sources are the Food and Agriculture Organization of the United Nations (FAO) and the World Bank (see Hussian 1986). The whole of this study has been conducted using data supplied by the Chinese statistical authorities, mostly the SSB. Agricultural data was initially compiled by the local collectives involving the production team, the brigade and the commune until the advent of the household responsibility system. This is essentially a self-reporting system with periodic checks for accuracy and reliability by the upper authorities (Hay 1979, p298). Such a system was based on the uniform rural institution and production organizations, especially on the principle of collective ownership. Since the system was designed to achieve full coverage to provide the information basis for the centralized planning system, there is no problem of sampling bias. The major source of non-sampling errors came from the reporting bias.

It is instructive to take grain production statistics as an example. On the one hand, there was a tendency of over-reporting by local officials who could thereby create personal political advantage; on the other hand, since state quotas were determined with reference mainly to the production level achieved (both for current and previous years), under-recording was a convenient way of retaining more grain either for team members' consumption or for selling to other marketing channels at a price premium. A general assessment on the size of reporting bias is very difficult for not only has it changed from time to time, but there are also great regional differences. Since the switch in policy to the households' responsibility system in 1978, the peasant household became the basic unit of production as well as accounting. To cope with this fundamental institutional change in rural areas, the former "self-reporting full coverage system" has been replaced by a stratified sampling survey system. Thus the man-made "response error" is being reduced, but replaced by potential sampling bias and aggregation errors. It is the task of the statistical authorities to monitor these procedures to ensure data of good quality.

In these circumstances analysts are left in a dilemma. Either they make use of published statistics including their inbuilt biases and thereby make precise but, possibly, poorly founded pronouncements, or they confine themselves to generalized qualitative statements. The former approach is used in this study and the appropriate qualifications are made to warn the reader where data are considered less than totally reliable.

The second set of issues concern the consistency over time and space of Chinese agricultural statistics (Stone 1982, pp205–207). China's statistical aggregation consists of two different systems: a central statistical system working at the national, provincial and county levels; and a statistical system inherent in various ministries which is also working at the national, provincial and county levels. For example, the Ministry of Agriculture is responsible for the collection and compilation of agricultural production data; it is the Ministry of Commerce who collects data on the purchasing of agricultural products and the selling of industrial inputs. Although in principle the State Statistical Bureau (SSB) is responsible for supervising the ministerial statistical work, in practice, the ministries define their own concepts, standards and collection procedures independently and in a way which is consistent with their administrative jurisdiction. Therefore, in using data from different sources, their distinctive features should be borne in mind. Another issue concerns the periodical fluctuations in the quality of Chinese statistical series. During the periods of political chaos (both the eras of Great Leap Forward and Cultural Revolution), the statistical system was badly hit like everything else. Consequently, not only were many detailed statistical materials not collected during these periods but some of the figures published were also modified according to the political taste prevailing, especially during the period of Great Leap Forward. Nevertheless, efforts have been made by the SSB to correct these man-made errors. A third area of concern are the variations in data quality between various economic sections as well as between different crops due to the difference in administration. Since the trade sector was totally controlled by the central government, data on agricultural trade is more accurate. It is considered that data for agricultural production is generally better than that of consumption. Within agricultural production, figures for grain and major cash crops are of better quality than those for vegetables, fruits and livestock production. Within the grain production statistics, the output is considered more accurate than that of either growing area or yield for the reasons explained in Section 4.2.2. From the consumption point of view, statistics on urban consumption are much more accurate than that of the self-subsistence dominated countryside (see Chapter Six).

The third set of issues concern the confusion over the definition, scope and method of compilation of agricultural statistics (Walker 1982). Since detailed explanations have been made available by the SSB during recent years, most of these problems have been resolved. For example, figures for grain output include soybean, potato and sweet potato but not that of Chinese yam and cassava; grain output is measured in terms of unmilled grain and that of consumption and trade in milled form; the conversion ratio from tubers to grain equivalent was 4:1 before 1963 and 5:1 since 1964 (see *TJNJ 1987*, p218) and so on. However, the question of whether the grain produce from peasant farmers' private plots was included in the production statistics or not has not been made clear officially. It is suggested that while most provinces have made estimations of production from private plots and added them to the aggregate grain output statistics, this item was always neglected by Jiangsu and Zejiang Provinces. It can only be hoped that this does not affect the data quality significantly.

Despite these problems, it is generally agreed that data from Chinese official sources is better in quality than that compiled by other institutions as well as the similar data for some other less developed countries (Hay 1979; Stone 1982). However, it should be clear from the foregoing that in order to use China's agricultural statistics properly, a deep understanding of both the basic features of the Chinese statistical system and technical aspects of data collection and compilation is very important. These cautions have been borne in mind whilst conducting the detailed analyses reported in Part II of this book.

Having outlined in this chapter the main features of Chinese agriculture and how it has evolved over recent decades, the next task is to summarize the development of policy towards the grain sector. This is the subject of the next chapter.

Chapter Two

The Historical Development of Chinese Agriculture: A Policy Review

2.1 Introduction

As a socialist country with a centrally planned economy, government policy has been of paramount importance for China's economic development. Since the late 1970s, Chinese agricultural economists have generally agreed that the historical development of the nation's agriculture (and for that matter the whole economy) since the founding of the PRC can be subdivided into two eras defined in socio-political terms. These were associated with changes in the Chinese leadership and political attitudes, and could be described as "realism and pragmatism" on the one hand and "ultra-leftism and romanticism" on the other. It is generally agreed that the economy performed more successfully in the first of these eras than during the other. Du (1984) and Huan (1985) were amongst the first to document this. However, the way in which these two socio-political eras influenced the design and implementation of development strategies and policies for agricultural development, and therefore the nature of the policy mechanisms and their effects on grain and agricultural development have not yet been systematically studied.

Amongst Western scientists with special interests in Chinese agriculture, most have identified the major policy issues and discussed their impacts on agricultural production in a piecemeal fashion. Examples of this approach are provided by Barker, Sinha and Rose (1982), OECD (1985) and Barnett (1981). More specifically, attention was mainly focussed on the evaluation of the effect of rural institutions and the agricultural planning system on the growth of output and productivity in agriculture since the founding of the PRC. For example, in approaching the supply response behaviour of Chinese agricultural producers, most analyses have taken a micro-economic approach, focussing on the production organization of collective agriculture. They usually came to similar conclusions as those of Schultz (1964) that peasant farmers are poor but efficient (see Parish and Whyte 1978; Rawski 1979). In his influential book published in 1983, Lardy shifted attention from the micro to macro level. He focussed on the impact of the broad socio-economic environment on agricultural production, that is, on "the large institutional setting in which collective farm units operate" (Lardy 1983, p18). From this work Lardy discerned two agricultural policy patterns which he called "indirect" and "direct planning". These ideas have enabled a better understanding of the relationship between policy and outcome. However, neither these micro nor macro approaches are appropriate or sufficient to describe fully and systematically the changes in the environment and how policy patterns affected the development of agricultural production. Moreover, it will be argued that agricultural planning has been only one of the dimensions in determining the decision-making and production management system of China's agriculture. Indeed the decision-making and management system itself is integrated with the policies on rural institutions and production organization on the one hand and the pricing and marketing system on the other. Thus, Lardy's description of policy patterns is regarded as too simplistic and thus not fully adequate for the task.

The purpose of this chapter is to identify the key policy elements relating to agricultural production in the last three decades. These elements are identified as: the rôle of agriculture in China's development, government policies on rural institutions and production organization, their policy towards agricultural planning and the management system, and the pricing and the marketing system. Each of these policy elements will be reviewed in turn. Their evolution since 1952 is then described. The outcome of this review is a set of theoretical hypotheses about the policy patterns and mechanisms under which Chinese agriculture progressed. These are presented and discussed in Chapter Three. The process of testing and validating these hypotheses, which is discussed in Chapters Four and Five respectively, will help answer why the Chinese agricultural sector developed along an erratic zig-zag course since the early 1950s.

2.2 The Rôle of Agriculture in China's Development

The development rôle for agriculture is a fundamental policy issue since it is directly linked with the objectives and development strategy for the economy as a whole. It has paramount importance in Chinese circumstances, for it deeply influences the detailed policy instruments employed and, consequently, the success or failure of agriculture as well as the whole economy.

2.2.1 The rôle of agriculture in theory

The contribution of agriculture to overall development has been intensively investigated by Western economists. Kuznets defined three types of contribution, the product, market and factor contributions. "If agriculture itself grows, it makes a product contribution; if it trades with others, it renders a market contribution; if it transfers resources to other sectors, these resources being productive factors, it makes a factor contribution" (Kuznets 1964, p114). Since its initial share in the economy is large in the take-off period, agriculture is a significant contributor to overall development. Agriculture makes a contribution to the economic growth by providing food and other basic needs for a growing urban labour force and at the same time provides a growing market for domestic manufactures of both consumer goods and industrial inputs used in agricultural production (Lardy 1983, pp12–13). The factor contribution of agriculture, that is the supply of labour and capital, occurs in two ways. First, capital is usually transferred through taxation. Second, savings which originate in the agricultural sector are used to finance growth of the non-agricultural sector (Kuznets 1964, pp114–115).

Since "every economy has an agricultural and non-agricultural sector and one of the most important aspects of development is the changing, complex but always intimate, relation between the two" (Youngston 1959, p284), a clear understanding of the relationship between agricultural and industrial-urban development becomes vital. Nicholls, whilst recognizing the complementary effects which agricultural progress and industrial development can contribute to each other, observed that agricultural progress is normally a prerequisite for industrial development. "Under all circumstances, increasing agricultural productivity makes important contributions to general economic development and, within considerable limits at least, it is one of the preconditions which must be established before a take-off into self-sustained economic growth becomes possible" (Nicholls 1964, p13).

Other economists have been concerned with the issue of balanced development. Lardy's view is that balanced development requires that agriculture is not regarded as simply a resource base which supplies labour, food and financial savings to the non-agricultural sector, but that it is a major component of a dynamic interactive growth process. In this way, the expansion of agricultural output "will promote overall economic growth and structural transformation and take full advantage of positive interactions between agriculture and other sectors" (Johnston and Kilby 1975, p133). However persuasive these theoretical ideas seemed, Nicholls (1964, p15) considered that for those countries which are seeking to launch rather than to sustain economic growth, the principle of balanced development is not easy to put into practice. In practice, the dilemma which faces underdeveloped countries is that the allocation of severely limited economic resources according to the criterion of balance may spread them so thin that productivity and income growth becomes imperceptible or even futile. Consequently hard choices must be made in the development plan and, according to Nicholls, the "major objective must be the concentration of scarce resources on certain strategic investments which will remove the most restrictive bottlenecks of the current situation" (Nicholls 1964, p15). This view gives rise to two groups of economists, one which emphasized agricultural development and the other emphasized industrialization.

The first group argued that efforts to increase agricultural production should receive priority. This is justified by the high demand for food and high marginal productivity of capital in agriculture (Leibenstein 1957, pp261–262). In the Indian case, the prerequisite for a successful development of the economy relies on substantial progress of the traditional sector of agriculture. "If one sector limits the growth of the other, it is more likely to be a case of agricultural growth limiting non-agricultural than vice versa" (Coale and Hoover 1958, pp120, 139). However, while recognizing the need for raising agricultural productivity, the second group of economists concluded that it can only be accomplished by giving priority to the industrialization programme. The only means to a cumulative improvement in agricultural productivity is a public policy "designed to make labour relatively scarce in agriculture by simultaneously shifting to a more mechanized and larger scale agriculture and encouraging a rapid rate of industrialization" (Higgins 1959, pp343, 459).

Chinese leaders certainly understood the rôle of agriculture and the importance of balanced development. There is evidence that these ideas had been identified earlier than in the West. For example, Mao Zedong was well aware of both the market and factor contributions of agriculture on his report *On Coalition Government*. "It is the peasants who contribute the main market for China's industry. Only they can supply foodstuffs and raw materials in great abundance and absorb manufactured goods in great quantities", and "in the future additional tens of millions of peasants will go to the cities and enter factories" (Mao Zedong 1943, p250). In regard to the balanced development, Mao criticized the idea of squeezing agriculture hard as "draining the pond to catch the fish" as early as 1942 (Mao Zedong 1942, p114; 1969, p633). In his article *On Ten Major Relationships*, Mao clearly stated that, if there is anyone who really wants to build up the industrial sector, he will have to encourage the development of agriculture; and similarly, someone who really wants an accelerated development of heavy industry, will have to develop agriculture and light industry first (Mao Zedong 1956, p722).

These ideas were shared by other senior Chinese leaders. For example, in his article entitled *The Development of Agriculture is of Paramount Importance*, Chen Yun, the chief architect of China's economic development in the 1950s, made it clear that since China is a country based on agriculture, it would be very difficult for the nation's industrial sector to be developed in the absence of corresponding development of the agricultural sector (Chen Yun 1951, p143).

However, as will be shown in the following sections of this chapter as well as in Chapter Five, the reality of historical development in the nation did not always coincide with what was described theoretically. The reasons for this are very complicated. Nevertheless, the most important contributing factors are summarized below.

First, after the founding of the PRC, China's development was significantly influenced by the traditional Soviet model of development, both in theory as well as in the management system adopted. During the first Five-Year Plan period (1953–57), Russian experts worked not only in many large investment projects providing technical guidance, but also in China's central administrative system, supervising the decision-making and economic management. These people were very influential even in organizations such as the State Planning Commission of China. Russia's model of development thus had a strong influence on China even after the specialist advisors retreated from China towards the end of the 1950s. Of course the Russian model of development itself was not without controversy. There was intensive debate in Russia during the 1920s and 1930s on industrialization as well as the Stalinist policy of "extraction" from agriculture. This was documented by Filtzer ed. (1980) and Hussain and Tribe (1981).

Second, is the crucial role played by Mao Zedong in all major decision-making concerning China's development for most of the time between the founding of the PRC to the mid 1970s. It is nowadays well known to the Chinese people that his views and his statements changed from time to time and his actions were not always consistent with his theoretical prescriptions. Generally before the founding of the PRC and during the early 1950s, Mao's attitude to China's development was rather pragmatic and cautious. Therefore although the Soviet model of development has played an influential rôle since the early 1950s, China's agricultural collectivization programme pursued during the first Five-Year Plan period was significantly different from that of Russia and it therefore had different consequences (see Nolan 1976). Unfortunately, Mao's personal attitude changed fundamentally in the late 1950s due to the internal political event of the so-called rightists attack on China's socialism and the communist party leadership in 1957. This culminated in the big political row with the Russian leaders which led China to be completely cut off from links with the Soviet Union as well as the whole Eastern Europe bloc. This is one of the main factors leading to the Great Leap Forward movement in 1958 which was subsequently followed by the Cultural Revolution from 1965 to 1977.

Third, as will be made clear in the following section, the choice of industrialization as the goal for China's development and, within this, the emphasis given to heavy industry was based on several important economic and socio-political considerations. The nation had suffered the impacts of war. At the time of the founding of the PRC it had a backward economy with a predominantly agrarian structure and weak industrial base biased toward consumer goods. In such circumstances launching an industrialization programme in favour of the development of heavy industry was, understandably, appealing for both the leading politicians as well as the general public. Painful lessons from the disruptions caused by such unbalanced development were yet to be learned.

2.2.2 The rôle of agriculture in practice

The goal set by the Central Committee of the Chinese Communist Party (CCP) and also contained in the Constitution of 1952 was "the gradual socialist industrialization of the country" (Kraus 1982, pp50,51). The major objectives were rapid economic development and transformation of China into a modern industrial power (Dernberger 1982, p67), thus the growth strategy placed priority on industry (Lardy 1983, p5).

There were four important economic, social, political and ideological considerations in choosing industrialization as the goal and giving priority to heavy industry.

First, industrialization was seen as the way of achieving economic independence. The backward economy with predominant agrarian and weak industry biased toward consumer goods (see Kraus 1982, p16 and *TJNJ 1983*, pp22,24) was seen as a semi-feudal and semi-colonial society. After the founding of the PRC, the new government believed that changing the backward, unbalanced structure of the economy, i.e. launching an industrialization programme with emphasis on producer goods, was of paramount importance. It was seen to be the only way of achieving economic independence to build on the political independence. It was also felt that political independence could not be strengthened without economic independence.

Second, heavy industry was defined by Li *et al.* (1957) and quoted from Tang (1968, p461) as the "modern industry which provides a material basis for the technical transformation of the national economy in the process of industrialization and for the modernization of national defence". Creation of a strong heavy industrial base was considered not only to be a precondition for a modern military establishment, but also the unique way to release the working class from hard physical labour and poor working conditions, which was another important political factor considered by CCP leaders.

Third, preference for the development of producer goods was one of the major strands of classical Marxian economic theory. After setting up a socialist system in China and the transition of the ownership of the means of production from private to public hands, the people were then seen as the "hosts" of the country and there would no longer be conflicts in terms of economic benefit among different groups in the society. Under these circumstances, the short-run and long-run benefits of the socialist system supposedly coincide with each other and squeezing agriculture for the purpose of industrialization was identified as contributing to the long-term benefit of the peasant farmers.

Fourth, market regulation systems based on private ownership were considered to be the fundamental characteristics of capitalist economy and a planned economic system built upon public ownership is the nature of socialist economy. This is one of the basic distinctions between socialist and capitalist social systems. Under the socialist system it was expected that the goal and strategy of development which fully reflects the long-term interests of Chinese people could only be effectively implemented under a planned system.

Many Western economists agree that the main features of the traditional Soviet model of development were repeated in China in the early 1950s. The rôle of agriculture in China's development revealed considerable similarities to that found in other communist countries (Eckstein 1961; Fei and Chiang 1966; Tang 1967, 1980; Ward 1980). That is, the agriculture sector served as a resource base to be "exploited". Thus the collectivization of agricultural production and the compulsory procurement of agricultural products by the state at well below market prices were the main features of the model (Dernberger 1982, p65). However, as pointed out by Tang (1968, p475) despite their common adherence to the traditional socialist model of development, quite different policies were pursued by the Russian and Chinese governments during their respective first Five-Year Plan periods. This was due, at least in part to the disparate initial conditions in the two countries. When the communists came to power in Russia, they "took over an economy with much more favourable preconditions for rapid industrial growth" than existed in China in the early 1950s (Hoeffding 1959, p39). Specifically, the per capita grain growing area, per capita grain output and per capita income in Russia in 1928 were, respectively, four times, more than twice and five times that of China in 1952 (Kraus 1982, p105). So, while the Russian government was able to extract agricultural surpluses for a long time to serve the cause of industrialization, China had first to find the ways and means for producing the necessary surplus. This is why in Tang's theory of development, Russia was classified as either Case I or II under which industrialization is not constrained by agricultural performance, but China was classified as Case III in which agriculture limits industrial growth, (Tang 1968, pp464–466). It also explains why China's development during the first Five-Year Plan period (1953–57) was described by Eckstein (1961, p508) as "a Socialist Strategy of economic development with local adaptation" and by Tang (1968, p504) "to be both extractive and developmental".

Other scientists, for example Aziz (1978) and Perkins (1980), argued that there were fundamental differences between Chinese development and that under the typical Soviet model. These writers felt that the objectives pursued and policies implemented coincided with the fundamental long-term interests of Chinese peasant farmers and thus the Chinese model had worked successfully. Such arguments were echoed by many others who referred to such matters as labour mobilization, rural employment and population control. They argued that agriculture was not squeezed by the government, and that the terms of intersectoral trade were continually improved in favour of agricultural development (Lardy 1976; Abbott 1977; Parish and Whyte 1978; Gek-boo Ng 1979; Rawski 1979).

However, given the dramatic way in which the major policies for agricultural development changed at several critical times over the last three decades, it is not possible to identify a "consistent pattern that can be interpreted as the Chinese model of development" (Dernberger 1982, p67). The rest of this chapter will show that there was neither a unique pattern in which the "traditional Soviet model of development" was applied nor a development path which could be described as "a successful example consistent with the fundamental long-run interests of the Chinese peasant farmers". In fact, as will become clear later in this chapter, changes in governmental policies for agricultural development could best be described as periodic oscillations between two policy patterns (see Figure 2.1 and Table 2.6). The first applied to most of the 1950s, early 1960s and the years since 1978. The second describes the experience during the Great Leap Forward in the late 1950s and the period of the Cultural Revolution.

2.3 Rural Institutions and Production Organization

Under the Chinese socialist system, the rural institutional framework is itself a product of policy (Griffin and Saith 1981, p121). The transition process of the rural institutions and production organization since the beginning of the 1950s started from land reform and evolved through the Mutual Aid Teams, the Elementary Producer's Co-operatives, the Advanced Producer's Co-operatives, the People's Communes, the system of three-level ownerships with the production team as the basic unit, and up-to-date the contracted responsibility system. Since this topic has been intensively investigated by many scientists (see Chao 1960; Walker 1965, 1968; Aziz 1978; Stavis 1982), only an outline of the basic features of each different stage is presented in this section, followed by a discussion about the nature and significance of the changes.

2.3.1 Land reform, 1950–52

To the CCP, Chinese agriculture was stagnant and depressed primarily because its institutional framework was feudalistic. In Marxist theory, the production relationships had impeded the development of productive forces (Wong 1973, p197). The land reform movement, which would destroy feudalism in China and return land to the tiller, was considered to be a revolution which would free Chinese peasants from the feudal yoke and create an essential condition for developing the forces of production and for China's industrialization (CHEC 1984, p8).

For the implementation of the land reform campaign, "The Agrarian Reform Law of the People's Republic of China" was promulgated by the central government on June 1950 (see Appendix B of Wong 1973, pp286–296). The distribution machinery, policy matters and process were cautiously designed and progressed (Luo 1985, Ch 2). After the land reform, China's 300 million peasants owning little or no land received, in total, more than 46 million hectares of cultivated land and other production facilities. The poor and middle peasants who accounted for 90% of the rural population and possessed less than 40% of arable land in the 1930s subsequently owned over 90% of the farmland. Correspondingly, the landlords and rich peasants who made up nearly 10% of the rural population and occupied more than 60% of the cultivable land in the 1930s retained only about 8% (Xue 1980, p19).

Land reform freed the Chinese peasants from feudal exploitation and thus raised their enthusiasm for farm production. However, the problems of small farm size, unequally distributed farming tools and other facilities, scarcity of credit and seed, and the implementation of a class policy, had adverse effects on further production growth and made some households unviable (Walker 1968, p399). More importantly, due to its low productivity and dispersed condition, the individual peasant economy was not considered to be able to meet the needs of industrialization (*CHEC* 1984, p16). This was confirmed by Luo (1985, p53), who noted that socialist industry could not carry on large scale planned production on the basis of a scattered, unplanned small peasant economy. The solution could only be found in the collectivization of agriculture.

2.3.2 Mutual Aid Teams (MATs), 1950–55

The establishment of producer Mutual Aid Teams was based on the traditional peasant custom of aiding each other in their labours. It also drew from the experience from the revolutionary base areas during the wartime. There were a variety of ways in which MATs operated. One system comprises short-term co-operation limited to cultivation and harvest seasons (seasonal MATs). Another is more permanent lasting the whole year (annual MATs). These were intended to remedy difficulties caused by the lack of productive means for example by shared use of draft animals and farming tools. Under MATs, individual ownership of land and other productive facilities was unchanged and member households received the products of their own farm. By the end of 1954, about 10 million MATs were established, accounting for 58% of peasant households (Kraus 1982, p76).

2.3.3 Elementary Agricultural Producers' Co-operatives (EAPCs), 1952–57

The Elementary Agricultural Producer's Co-operatives, which pooled land, labour and capital in units of 20 to 25 households, were initiated in 1952 (see Table 2.1). They were frequently formed along the existing kinship ties at the village level. Membership was voluntary and workers received income based on the land, capital and labour contributions. About 60–70% of the total income was distributed on the basis of labour input and 30–40% of the income was distributed as dividends or rent for land and other productive means contributed by individual members. Management of the EAPCs was the responsibility of a committee elected by all the members annually (Aziz 1978, p12). In addition, all households retained private plots for their own use (Dernberger 1982, p72). At the end of 1955, 14.2% of the peasant households worked under the EAPCs (Table 2.1).

2.3.4 Advanced Agricultural Producers' Co-operatives (AAPCs), 1956–57

The transition from semi-socialist co-operatives to socialist collectives was proposed by Mao Zedong in his report entitled *On the Cooperative Movement in Agriculture (*Mao 1955). A surge of co-operation ensued. The launch of the Advanced Agricultural Producers Co-operatives in 1956 coincided with similar movements for the socialist transformation of capitalist industry and commerce in the city and handicraft industries in urban and rural areas. This movement was accompanied by a political campaign to educate peasants (Luo 1985, pp73–74). The AAPCs became full socialist organizations because land and the other major means of production were collectively owned and income was distributed according to the labour contributed and nothing else (CHEC 1984, p18). The main features of the AAPCs were outlined by Walker (1968, pp402–404) as,

1) Collectivization of land without compensation.
2) Draft animals and large farm implements became collective property at agreed prices.
3) Private plots (maximum 5% of the arable area) were allocated to the members for their own use.
4) Personal income from the collectives was entirely derived from labour according to the norm set by the AAPCs.
5) The AAPCs were between 100 and 300 households in size. It was further divided into subdivisions which were equivalent to the EAPCs in scale. Each of the sub-units was a permanent unit with its own labour, land and other facilities.
6) The principles of "voluntariness" and "mutual benefit" were stressed.
7) The initial production responsibility system was implemented within the AAPCs.

The progress of agricultural reorganization during 1952–57 is summarized in Table 2.1.

2.3.5 The People's Communes, 1958–60

In 1958, a most radical reorganization of agriculture was carried out by the government. More than 740,000 AAPCs were merged into about 23,000 People's Communes, with an average of over 5,440 households in each commune (*TJNJ 1983*, p147). Three main features were clarified by Walker (1968, pp442–443).

(1) A commune was an administrative unit of government as well as a comprehensive planning and management organization. The commune authority itself became the main unit of ownership and distribution. This is why Wiles (1962, p44) regarded it as an all-embracing rural institution.

(2) Distribution according to labour was substituted by distribution partly on need and partly on work accomplished.

(3) No private sector was allowed by the Commune's regulation and it was often totally abolished in practice.

Disastrous results were created from the excessive commune size, a militaristic style of labour allocation, poor incentives and abolition of the private sector during the period of 1958–60 (Stavis 1982, p91). These were further exacerbated by severe natural calamaties during 1959–61. As a result, famine spread over the country. The situation forced the Chinese government to retreat.

2.3.6 The system of "three-level ownerships with the team as the basic unit"

The retreat period 1961–65 The major reorganization of the commune system was started after a letter was issued by the CCP's Central Committee in November 1960 (usually known as "12 rules"). A system of three-level ownership was set up. At the highest level was the commune itself, this was subdivided into production brigades and, in turn, the lowest level was the production team. The production team was defined as the basic unit which, because of its much smaller size, was easier to control and more sensitive to the needs and capabilities of its members. These policy measures were further confirmed and detailed by the *Rules and Regulations for the*

Table 2.1 Agricultural reorganization, 1952–57

Year	Independent peasant households	Mutual Aid Teams			Agricultural Producers' Co-operatives		
		Total	Seasonal	Annual	Total	Semi-Socialist	Socialist AAPCs
	percent of peasant households						
1952	60.0	39.9	29.8	10.1	0.1	0.1	–
1953	60.5	39.3	27.8	11.5	0.2	0.2	–
1954	39.7	58.3	32.2	26.2	2.0	2.0	–
1955	35.1	50.7	23.1	27.6	14.2	14.2	–
1956	3.7	–	–	–	96.3	8.5	87.8
1957	2.0	–	–	–	98.0	2.0	96.0

Source: from Kraus 1982, p102 (according to the figures from the SSB)

Works of the People's Communes (also known as the *Sixty Articles*) promulgated in January 1961. Since 1961, the production team has thus operated as the basic planning, management, accounting and income distribution unit. Private plots and free markets were reinstated. Furthermore, the household contracted responsibility system was being utilized in some localities, the famous story in Anhui countryside was presented by Luo (1985, pp132–134). Thus, during this period, as confirmed by Tao Zhu (one of the senior leaders of the CCP), it "was like the former collective, particularly in its system of responsibility and distribution" (quoted from Walker 1968, p447) and by Aziz (1978, p16) that "in addition to restrictions on the buying and selling of land, this approach would have meant a restricted return to private farming".

The Cultural Revolution 1966–77 During this period, political propaganda distributed by the Communist Party argued that China's countryside was threatened by the resurgence of the landlord and bourgeois classes and there existed a sharp struggle between "two classes, two roads and two lines" in the rural economy. Other influential slogans proclaimed "we must keep political, instead of economic accounts" and "better to grow socialist weeds rather than capitalist shoots" (Luo 1985, pp89–90). Under this political environment, rural institutions and production organization suffered again in three ways.

Transition in poverty It was argued that the poorer the people, the more revolutionary they are. So transition to communism should be carried out when people are still poor. As a consequence, 80,956 communes, 652,000 brigades and 5,643,000 production teams existing in 1963 were amalgamated by 1970 into 51,478, 643,000 and 4,564,000 units respectively. Average household numbers per commune increased from 13,424 to 15,178 over the same period (*TJNJ 1983*, p147). The proportion of communes with brigades as the basic accounting unit rose from 5% in 1962 to 14% in 1970.

Cutting off the tail of capitalism For most times during this period, private plots and household sidelines were regarded as "the tail of capitalism" and had to be "cut off". For example, in some areas, raising three chickens was allowed for each household and any more than that was capitalism. In some other places, it was decreed that the number of chickens should not exceed the number of people in the household, otherwise the extra number had to be confiscated.

Banning county fairs In order to eliminate thoroughly the rural fairs, a large scale campaign to "revolutionize commerce" was launched. In some instances armed men were sent to mop up the fairs.

2.3.7 The contracted responsibility system since 1978

The turning point of Chinese agricultural policy after the Cultural Revolution was marked by the Third Session of the Eleventh Central Committee of the CCP, which was held at the end of 1978. Since this congress, the ultra-left tendency was criticized and major policy changes were once again implemented.

The core of the new policies is the promotion of the contracted production responsibility system (PRS). It is the means "to put to an end... the practice of "eating from the common pot", to drive away the spectre of egalitarianism and establish an appropriate balance between personal responsibility and social welfare" (Luo 1985, pp126–128). The general trend of progress in the responsibility system since 1978 is from work-contract to output-contract on the one hand, and from group contract to household contract on the other. The work-contract system was based on contracting farm-tasks without direct linkage to the final outcome. Under the household contracted responsibility system, while the collective ownership is still upheld, land is contracted to the peasants for cultivating (Du 1984, p19), and the obligations of the peasant household to the state and the collective "amounts to nothing but a lump-sum tax" (Kueh 1984, p355). The switch to the output-contract system mostly took place in 1983 and 1984. By this latter date practically all production teams had converted to the output-contract system and this accounted for all but 2% of peasant households. There is little doubt that this system is very popular with peasant farmers because of its simplicity (Luo 1985, p131).

In addition to the implementation of the production responsibility system, development of the household sidelines and free market trade were also encouraged by the government (Schran 1982, p433). Since 1978 the ceiling for the land in private use (including private fodder land) as a proportion of the total farmland has been raised from 5.7% to 15% (*CHEC* 1984, p258). From 1978 to 1984, the number of free markets for agricultural and sideline products increased from 33,302 to 56,500 and the value of transactions was estimated to have risen from 12.5 to 47.6 billion yuan over the same period (*China's Rural Statistical Yearbook 1985* (*Zhongguo Nongcun Tongji Nianjian 1985*, hereafter *NCTJNJ*), p136).

2.3.8 Summary

Since the land reforms of the early 1950s, although the form of agricultural organizations changed from time to time and took different names for different periods, two uniform patterns of organization can be identified according to their basic features.

The common features for the periods of 1952–57, 1961–65 and 1978–84 are: (1) the important rôle of the peasant households in the development of agricultural production and the adoption of responsibility system inside the basic production units; (2) the principles of voluntary participation and mutual benefit; (3) the tight link between contribution and income distribution; (4) the encouragement of household sidelines, free markets and private plots production. This pattern of institutions and production organization will henceforth be referred to as agricultural co-operatives.

The basic features for the periods of 1958–60 and 1966–77 are: (1) a high degree of influence from the political environment; (2) a lack of devolved responsibility inside the production units, and particularly the demand of a significant decision-making rôle for peasant households in either production or distribution; (3) the large scale of the basic production organizations with "everybody eating from the same public pot"; (4) prohibition or constraints on private plots, household sidelines and the free market. This pattern will be referred in the rest of this book as collectives.

2.4 Agricultural Decision-Making and Management System

2.4.1 The historical evolution of the planning and management system

Before 1956 Although the administrative planning system was set up in the early 1950s (Kraus 1982, p41), for several reasons, direct control of agricultural production by the central government was not carried out and development was thus guided by the free market.

Creation of co-operatives means pooling labour and other productive factors and sharing output accordingly. This was based initially on private ownership by peasants. Under such circumstances, "individual peasants cannot be directly controlled by the state plan. They formulate their production plans according to their own requirements and particularities" (Teng 1955, pp9–13).

The state controlled only 10% of purchases from the agricultural sector in 1950 and less than 25% in 1951. The absence of an established state commercial network and other relevant facilities made direct control impossible (Perkins 1966, p31).

More than 120 million peasant households and the nature of agricultural production made planning much more complicated than the planning authority could manage. In this situation, the formulated agricultural production plan is necessarily imprecise and its specification is scattered and regional in nature (Yen 1977, p35). The situation was thus described by Lardy as, "to meet the objectives of increasing farm output and raising the marketed share, the state relied primarily on price incentives" (Lardy 1983, p30).

Between 1956 and 1957 After the setting up of the AAPCs, direct implementation targets drawn up by the central planning authorities were disaggregated and passed down the administrative hierarchy, and finally to the AAPCs (Lardy 1983, p37). This direct plan was mechanically accepted by most of the basic production units (Walker 1968, p424). However, it was ignored by some others because of the poor quality of the plan. Conflicts arose between fulfillment of the state's objective to maximize the rate of production growth and the AAPCs' goal to maximize income (Perkins 1968, pp62–66).

When it was proved to be impossible, direct implementation of the production plan was quickly abandoned and steps were taken by the government to change the situation. In 1957, stress was laid on the importance of flexibility at the lower levels in adjusting targets according to local conditions; the number of targets set by the administrative authorities was reduced; and these plans were only to be used as references and guidance for the co-operatives (Perkins 1968, pp60–66).

1958–60 When the People's Communes were set up in 1958, direct control of production, exchange, consumption and accumulation under the state plan was stressed by the government (Wang 1959, p16). Decision-making and management within the communes was concentrated in the hands of local cadres. Peasant farmers' rights to determine their own pattern of cropping and allocation of output, after meeting state procurement targets, were eliminated (Lardy 1983, p41). The situation was made even worse by the huge pressure put on commune leaders to achieve spectacular results, which made it safer for local cadres to deviate towards "greater centralization and reduction of individual control" (Perkins 1968, p85).

1961–65 The economic crisis and widespread famine throughout the country in 1959–61 brought the Chinese leaders back to reality. By 1961 the level of accounting and production decision-making was moved from the brigade to the team level which were of a size equivalent to the EAPCs. Production plans were no longer implemented below the county level (Lardy 1983, p44). In addition, the principle that any central directive has to be applied with considerable flexibility was once again emphasized and more diversified cropping systems for the better use of agricultural resources were encouraged.

1966–77 During the political chaos caused by the Cultural Revolution, agricultural production was once again planned by the administrative system who passed the plans directly to the basic production units for execution. It was argued that direct state control of agriculture was necessary to prevent a spontaneous tendency towards capitalism (Wang *et al.* 1965, p36). As a result, production teams had no scope to make decisions, they had simply to obey orders. In practice, "the emphasis was on unified action – unified plans for planting, unified working hours, unified requirements for ploughing and unified specification for farm work" (Luo 1985, pp107–108). Furthermore, the size of basic production units was once again expanded and indiscriminate transfer of resources from the basic production units prevailed (Luo 1985, pp92–94). Finally, the strategy of "take grain as the link" designed to achieve grain self-sufficiency during the Great Leap Forward was resurrected by the government. The result was irrational cropping patterns, stagnation or decline in cash crop production, and an increase in production costs (Lardy 1983, pp47–88).

1978–84 Since 1978, several important policy measures have been pursued to improve the decision-making and management in agricultural production. These are, to respect and safeguard the basic production unit's right of ownership (CHEC 1984, p255); to recognize its freedom in decision-making and production management (Schran 1982, p431); and to encourage the development of a diversified rural economy (Kueh 1984, p358).

With the implementation of the household production responsibility system, peasant households became the basic production unit (Huan 1985, p15). As a consequence, centralized management is integrated with decentralized management (Du 1984, p19). On the one hand, peasant farmers take on the duty for production decisions and management within their contract with the production team; on the other hand, the production team co-ordinates farming activities not only by means of the contracts with producers, but also with the large public facilities (An 1983, p6). In this way, the production plan issued by the central government is a guide for the grassroot production organizations linked through the hierarchical administrative system. Meanwhile, the number of planned targets was reduced from 21 in 1978 to 13 in 1982 (Wu 1983, p4).

2.4.2 Summary

This review reveals that although the agricultural planning and management system changed from time to time after the land reform, two basic patterns can be identified. One is the system pursued during the first half of the 1950s, 1957, 1961–65 and 1978–84. The common features for these periods are: (1) The co-operative ownership and its responsibility for decision-making and production management was respected and protected. (2) The complex nature of agricultural production was allowed for as considerable flexibility permitted local conditions to be taken into account. (3) Agricultural production plans from the central government were not allowed to be transmitted down to and directly implemented by the basic production units. In practice, production plans functioned as general guidelines. Furthermore, the number of targets in the production plan was reduced. (4) Pricing and marketing policies (which will be reviewed in the next section) were used as the main policy instruments for inducing the development of agricultural production. This system will be referred to as "indirect" control.

The other pattern is the system used during the periods 1956, 1958–60 and 1966–77. The basic features for these periods are, (1) The decision-making and production management system were under great political pressure. The directives from the centre were to be executed mechanically and quite often were further exaggerated by the local cadres. (2) The importance of agricultural production was not emphasized, the ownership and freedom of the basic production units were neglected and violated. (3) Not only were the production plans passed down from the top to the bottom and directly implemented by the basic production units, unified actions were also asked by the top leaders and pursued by the local cadres. (4) Local grain self-sufficiency was emphasized and peasants were urged to concentrate agricultural resources on growing grain crops. This pattern of decision-making and management will be referred to as "direct" control.

It should be noted that the switches between the "direct control" and "indirect control" systems always coincided with shifts in the political environment (Perkins 1966, p85 and Du 1984). Furthermore, the oscillations were consistent with the changes in production organization and rural institutions. The reason for this has been mentioned by Chao (1960, p152), Yen (1977, p35) and Liu (1977, p180). That is a transition in production organization is a precondition for a change in system of control. Both of these elements are seen as necessary to serve the goals and strategy of development.

2.5 The Pricing and Marketing System

2.5.1 Agricultural tax

Before the compulsory procurement system was introduced for major agricultural products, agricultural taxation was the main method used by the government to draw income from the agricultural sector. The rate of the tax was based on the normal yield of the land, that is the amount the land should produce under normal climatic conditions. The tax was generally paid in kind (Donnithorne 1970, pp4–5; Gek-boo Ng 1979, p75). In 1955, about 85% of the tax was paid in grains and 8% in cotton (Chao 1960, p200). The unit directly responsible for payment of the tax changed with the system of production organization. The superior level was always held responsible for ensuring that the lower units fulfilled their tax obligations (Donnithorne 1970, p5). In quantity terms, the tax averaged 18.6 million tonnes of grain per year during 1951–57, and was reduced to about 13 million tonnes in 1978 and less than 11 million tonnes in 1979. The reductions resulted from cuts in the tax rate or exemption for the poorest region (Lardy 1983, p103). The combination of a declining quantity of tax revenue and the increase in farm output meant that the real tax rate fell from 12% of the agricultural output in 1952 to 4.4% in 1978 (Keng 1981, p17). The tax payment in value terms (in current prices) is shown in Table 2.2.

2.5.2 The evolution of pricing and marketing systems

The agricultural production tax became a less and less important channel with the passage of time and the main impact of government intervention switched to quantities purchased and their price (Walker 1984, p45). Thus it is important to consider the development of the pricing and marketing system.

The basic features of China's development in the early 1950s have been referred to above. The most serious problem was the growth in grain production and its availability to the government. In 1953, which marked the beginning of the first Five Year Plan period, the urban population increased from 71.6 (1952) to 78.3 million

Table 2.2 Agricultural tax and its proportion of the state financial income, for selected years

	1950	1952	1957	1962	1965	1970	1978	1983
Agricultural tax billion yuan	1.9	2.7	3.0	2.3	2.6	3.2	2.8	3.3
% of the state financial income	29.4	14.7	9.6	7.3	5.4	4.8	2.5	2.6

Sources: Data for 1950–78 is from *TJNJ 1983*, pp446–447
Data for 1983 is from *TJNJ 1985*, pp523–524.

persons, a net growth of 6.6 million people; capital investment rose 103% above the 1952 level and foreign exports were 28% higher (*TJNJ 1983*, pp103, 323, 420). As a result, a disequilibrium between the state's acquisition and disbursement of grain arose amounting to 3.5 million tonnes during the 1952–53 grain year (July–June) (Wu 1957). Given a high income elasticity of demand for grain and a widened gap between the state's purchasing price and the price on the free market, the peasant's willingness to sell grain was reduced (Walker 1984, p43). The Chinese government faced a dilemma at the end of 1953, they had either to cut back the investment programme and slow down the pace of industrialization or abandon the market mechanism and take radical centralization measures to secure the necessary grain deliveries to the state. The government chose the latter option (Chen 1952, p160 and 1953a, p210). It was believed that, although China's grain production level was low, there could be enough both to meet domestic needs and to provide an export surplus if grain was rationally distributed (Chen 1954, pp254–263). As a consequence, a compulsory quota system for grain was imposed in December 1953 and followed by similar schemes for cotton, oil-bearing crops and other major agricultural products (Chen 1955, p274).

1953–57 Under the compulsory quotas system of grain, the market was monopolized by the state. Peasant households deemed to have surplus grain were required to sell specified amounts to the state at prices fixed by the state (Walker 1984, p43). The state increased the grain taken from the peasants during 1953–54 (see *TJNJ 1983*, p393). The tension created between government and peasants resulted in "a great many people not welcoming us" (Mao 1969, p229). In order to stabilize peasants' obligations and to minimize the disincentive effect, the "three-fixed" policy was adopted (Donnithorne 1970, p6). This meant, first, that a standard production quota was allocated to each unit of land (fixed production). Second, quotas were set for compulsory sales to the state (fixed purchase). Third, grain rations were fixed for selling back to the grain-deficient peasants (fixed sales). These were mostly to peasants who grew non-grain crops. This system was to remain unchanged at the level of 1955 grain year for the three years afterwards (Chen 1955, pp276–277).

Given the fall in the price ratio between agricultural and industrial products from the 1930s to the end of the 1940s (Perkins 1966, p78), the purchasing price for grain was raised about 20% between 1950 and 1952 (Chu 1958, p34). From 1952 to 1957, the price index for agricultural products and grain crops was increased by 20.2% and 16.5% respectively (in 1952 constant prices, see Table 2.3). The intention was to encourage peasant farmers to rehabilitate and develop agricultural production as well as deliver products to the state.

The implementation of the compulsory procurement system for grain was generally successful during 1953–57. Within this period, an annual average of 48.7 million tonnes of unmilled grain was purchased by the government representing 27% of the total output. After the deduction of grain sold back to the countryside, net purchasing averaged 33.2 million tonnes per annum which was 18.4% of the output (Table 2.4). Thus it was considered critical for the Chinese government to

Table 2.3 Purchasing price index for agricultural and sideline products (1952 = 100)

	General index	In which			
		Index for grain products	Index for cash crops products	Index for livestock and poultry products	Index for other agricultural and sideline products
1952	100.0	100.0	100.0	100.0	100.0
1957	120.2	116.5	111.9	137.7	130.9
1960	129.4	125.0	118.4	154.0	145.8
1965	154.5	170.4	136.7	181.7	156.5
1977	172.0	211.7	149.1	190.0	167.6
1983	264.2	379.0	241.6	254.2	234.0
Average annual increase for these periods (%)					
1952–57	3.8	3.1	2.3	6.6	5.5
1958–60	2.4	2.4	1.9	3.8	4.5
1961–65	3.6	6.4	2.9	3.4	1.4
1966–77	0.9	1.8	0.7	0.4	0.6
1978–83	7.4	10.2	8.4	5.0	5.7

Note: The purchasing price indices in this table are based on the real prices offered by the official marketing system. They are weighted price indices covering compulsory procurement prices, prices for above-quota purchases and negotiated prices.
Source: Data is derived and calculated from *ZCWTZ 1952–1983*, p402.

carry out its industrialization programme during the first Five-Year Plan period. However, the implementation of the government's policy faced great difficulty. The tension was always high over the grain procurement during this period, "with everyone talking about grain; and households all talking about central purchasing", moreover "old women blocked the road and would not allow the grain to be taken away" (Mao 1967, pp3–7, 160). As a result, the government lowered procurement targets to reduce the peasants' opposition. The proportion of purchasing and net purchasing was reduced from 28.4% and 21.5% in 1953 to 24.6% and 17.4% in 1957 respectively (Table 2.4). The government's grain balance sheet was in a state of "tense balance" throughout the whole period (Sun 1958, pp24–27). A review of the effects of controls on cotton, oil-bearing crops and other agricultural products can be seen from Chen (1953b, 1953c, 1956, 1957a) and Perkins (1966, pp53–54).

1958–60 With the anti-rightists and socialist education campaigns launched in the second half of 1957, and Great Leap Forward and People's Commune

Table 2.4 Per capita grain output and procurement

	Per capita grain output (kg)	Total procurement		In which: Net procurement	
		Amount (million tonnes)	% of total output	Amount (million tonnes)	% of total output
	(a)	(b)	(c)	(d)	(e)
1953	283.7	47.46	28.4	35.89	21.5
1957	301.7	48.04	24.6	33.87	17.4
1960	216.7	51.05	35.6	30.90	21.5
1965	268.2	48.69	25.0	33.60	17.3
1977	297.7	56.62	20.0	37.56	13.3
1984	393.6	141.69	23.2	94.61	23.2
Annual average for:					
1953-57	294.6	48.70	27.0	33.24	18.4
1958-60	257.7	59.07	34.9	40.06	23.5
1961-65	248.4	43.74	25.5	29.18	17.0
1966-77	289.1	53.50	21.7	39.22	15.9
1978-84	346.6	90.69	25.7	60.86	17.2

Notes: (1) Annual population is the total number at the end of calendar years; Annual grain output, total and net procurements are for "agricultural production year"(April-March).
(2) Data for grain output and procurement are for unmilled grain. Net procurement is the total procurement less that sold back to rural areas.

Sources: Column (a) is derived and calculated from *TJNJ 1983*, pp103, 393; *TJNJ 1985*, p185, 482.
Columns (b), (c), (d) and (e) are derived and calculated from *TJNJ 1983*, p393, and *TJNJ 1985*, p482.

movements in 1958, the government's position in controlling grain distribution and consumption was further strengthened. The "wind of exaggeration" brought severe pressure on institutions at all levels to report drastically inflated output figures. There was increased pressure to sell more grain to the state (Chen 1957b, pp60–65; Donnithorne 1970, p7). During this period, the compulsory purchasing of grain was regarded as a serious political issue, testing "who genuinely supports socialism and who genuinely loves his country" (*The People's Daily* 1957, quoted from Walker 1984, p66). "It was a period of fierce political struggle" (Walker 1984, p65), and peasant farmers were undoubtedly the losers for they had no effective means to safeguard their survival under the given institutions, management system and political pressures.

During this period the free market was closed, grain self-sufficiency in all agricultural areas was under stress, and the state reduced the amount of grain sold back to the rural areas. As a result, 59 million tonnes of grain in total and 40 million tonnes net was extracted when the average per capita unmilled grain output was only about 257 kg. The 34.9% of total and 23.5% of net procurement rates were higher than at any other period since the 1950s (Table 2.4). The most serious situation was created in 1960, in which 35.6% and 21.5% of the total and net procurement rates were achieved given the lowest per capita grain output of 216.7 kg. It was in this year that a negative natural growth rate of population of 4.5% was registered, corresponding to a 10 million net reduction in total population (*TJNJ 1983*, pp103–105). These statistics scarcely do justice to the scale of human suffering endured at this time.

1961–65 At the beginning of 1961, the government was forced to shift its development strategy. Policy measures were taken during this period for restoring the agricultural situation which had been set back ten years by the Great Leap Forward and People's Commune movements (Luo 1985, p86). To coincide with the changes in the production organization and management system, a new policy for improving pricing and marketing system emerged with the following three main features.

First, the purchasing price for grain was increased 25% in 1961 (Liu 1980, p28). The internal price ratio between grain and other agricultural products was also adjusted to induce production onto the right track (*TJNJ 1983*, pp478–482). Between 1960 and 1965, the purchasing price index for agricultural and grain products was increased 15.1% and 45.4% respectively (Table 2.3).

Second, the government released direct market controls. One of the lessons for Chinese leaders was that to control agricultural product markets, looser is better than tighter (Chen 1961, pp146–148). The "three-fixed" policy was rescinded during this period (Chen and Ridley 1969, pp81–89). The products directly controlled at the central level were reduced in number and the free market was once again opened.

Third, grain purchasing rates were reduced. The average total grain purchased fell from 59.1 million tonnes over 1958–60 to 43.7 million tonnes during 1961–65. Net grain purchased fell from 40.1 to 29.2 million tonnes per annum over the same period. Thus, the total and net purchasing ratios declined from 34.9% and 23.5% to 25.5% and 17% respectively (Table 2.4). Meanwhile, a large amount of grain was imported annually to improve grain distribution and consumption conditions (see *TJNJ 1983*, p438).

1966–77 During the chaotic period of the Cultural Revolution, a sharp political struggle between "two classes, two roads and two lines" was fought. At this time grain sales to the state once again became a political issue. At each harvesting season, working groups were sent by the People's Commune or higher authorities to the basic production units to collect the "patriotic" grain. Not only was free market trading prohibited, but also basic production units which did not fulfil the compulsory quotas were not allowed to allocate grain and other products to their members. Meanwhile the procurement price level for grain and other major crops were hardly changed throughout the whole period. As shown by Table 2.3, the purchasing price index for agricultural products was increased by 0.9% per annum only. For grain products, after the increase made in 1966, the purchasing price declined until the year 1970 (*ZCWTZ 1952–1983*, p402). As a result, the price gap between agricultural and industrial products widened considerably, from about 20% in 1952 to more than 50% for most of this period (Chen 1983a, p95). During this period, 53.5 million tonnes of grain were collected by the government annually, and 39.22 million tonnes per annum were made available for distribution outside the rural areas (Table 2.4).

1978–84 The decision to increase substantially the purchasing price was made by the Third Session of the Eleventh Central Committee of the CCP. The state compulsory purchase price for grain was raised by 20% and the premium paid for above-quota purchases was raised to 50%. Furthermore, extra purchasing with negotiated prices was introduced for grain and many other agricultural products. After fulfilling the tax levies and the targets for quota and above-quota purchases, producers could either negotiate with the official marketing agency for selling additional produce or trade it in the free market directly. Meanwhile, price was used as a major policy instrument to influence the composition of output (Lardy 1983, pp89–92; Luo 1985, pp103–104). Also the number of targets for agricultural products directly controlled by the central authorities was reduced from 31 in 1978 to 13 in 1982 (Wu 1983, p4).

Significant real price increases were achieved during this period. As shown in Table 2.5, during 1978–84, the composite average purchasing prices for milled grain, ginned cotton and edible oils (from oil-bearing crops) were increased by 50%, and the prices for sugarcane and sugarbeet were raised by 67% and 30%

Table 2.5 Composite average purchase prices for major farming products (yuan/tonne)

	1978	1984	% increase during 1978-84
Milled grain	263.4	395	50
In which: compulsory quotas	256	318.8	24.5
Ginned cotton	2,278	3,418	50
Edible oils	1,746.4	2,624.4	50
In which: compulsory quotas	1,675.8	2,289.8	36.6
Sugarcane	36.2	60.6	67.4
Sugarbeet	60.5	78.6	29.9

Notes: (1) Prices in this table are the current prices.
(2) For the Chinese statistical publications, composite average price usually is the weighted average price after taking account of the differences in standard, grade and quality of a commodity. One more dimension had been considered in constructing this table, i.e. the weighted average for the proportional composition with compulsory quota price, above-quota premium and the negotiated price of the commodity.

Source: From *NCTJNJ 1985*, p152.

respectively. The purchasing price index (in 1952 constant prices) for agricultural products was increased from 172% in 1977 to 264.2% in 1983, with a growth rate of 7.4% per annum. Within this, the index for grain products rose from 211.7% to 379% in the same period. This represented an average annual increase of 10.2% (Table 2.3).

During the first few years of this period, although grain output grew rapidly in response to incentives created by the government, state grain procurement was cautiously controlled. Not only was less grain acquired as the tax was reduced from 19.1 million tonnes, 10.5% of the output in 1953–57, to around 13.8 million tonnes, 4.5% of the output in 1977–80 (Walker 1984, p149), but also the net grain purchased by the state was restricted at 15% until 1982 (see *TJNJ 1983,* p393). Since the per capita grain production in 1977 was less than 20 years previously (1957), and more than 100 million peasants were still suffering from "insufficient foodgrains" (CCP Central Committee 1978, p151), there was still considerable incentive to encourage peasant farmers to increase production further. During 1983–84, the combination of a continuing boom in production and improved incentives resulted in a large increase in state grain purchasing. In 1984, total grain purchased by the state amounted to 142 million tonnes, nearly 35% of the total output (Table 2.4). For the first time the government had to worry about such problems as insufficient storage and marketing facilities for the very large crop rather than the more usual problems of trying to obtain sufficient grain for the state.

2.5.3 Summary

From the brief review in this section, two patterns of pricing and marketing can be identified. For the periods of 1953–57, 1961–65 and 1978–84, price was used as a major policy instrument to encourage peasant farmers both to produce and to sell. The price index for agriculture and grain products was increased by 3.8% and 3.1% per annum in 1952–57, 3.6% and 6.4% during 1961–65, and 7.4% and 10.2% for 1978–83 respectively (Table 2.3); the free market was opened and the level of direct control by the central authorities relaxed. Certain bargaining power in procurement was placed in the peasant household's hand, and government targets for grain procurement were designed and implemented with caution, whether consciously or involuntarily. This pattern is termed "looser control and improved prices". During the periods of 1958–60 and 1966–77, purchasing prices for grain and other agricultural products stagnated, and the price gap between agricultural and industrial goods widened. The free market was prohibited and more items of agricultural products were directly controlled at the central level. Political pressures were imposed on grain procurement, and a high proportion of the crop was officially purchased given the low level of per capita grain production. Little consideration was given to the resulting disincentive effects on peasant farmers. This pattern may be summarized as "tight control and poor farm prices".

This brief review of historical development has attempted to show that it was the combination of price measures and marketing procedures which were the most important instruments through which the government extracted economic resources from agriculture to be used for overall development. In this process it was necessary for the pricing measures to be backed up by appropriate marketing procedures. It is also clear that policy on pricing and marketing systems was integrated with the changes in production institutions and management systems. Without the collectivization movement and direct control system, it would not have been possible for the Draconian act of "draining the pool to catch the fish" (Mao 1969, p633) to have been carried out during the eras of the Great Leap Forward and the Cultural Revolution.

2.6 Theoretical Hypothesis of the Policy Patterns and Mechanisms

The review of the historical development of China's agricultural production in Chapter One (see section 1.3) revealed that the development of China's grain production followed an extremely erratic course since the early 1950s. As there was no coincidental change in resource availability or technological improvement, and as climatic factors were unlikely to have been large enough to explain the change in output, shifts in government policies are suspected to be the main factors explaining these development patterns. Based on the policy review described above, the principal hypothesis of this book is that the development of China's grain and agricultural production was significantly influenced by three major policy ingredients: (i) rural institutions and production organization; (ii) the decision-making structure and production management system; and (iii) the agricultural pricing and marketing system. These main policy factors acted in a concerted way and they were also integrated with and determined by the agricultural development strategy. This in turn was dependent on the strategy of national economic development, and, above all, the political environment. These relationships are summarized in Chart 2.1. Within this framework it is proposed that during the last three decades since the founding of the PRC, there has been a fundamental dichotomy in the political economy of Chinese agriculture which can largely explain the course of her agricultural development. The nature of this dichotomy has been described in this chapter and is summarized in Chart 2.2.

Thus during periods when the "ultra-left" ideology was pursued by the leadership, the mode of running the country could be called "romanticism". For political purposes, rapid development of the national economy is required, necessitating an emphasis on production goods at the expense of consumer goods. The consequence is a rapid and accelerating development of heavy industry. To realize this strategy, cheap economic resources, including labour, consumer goods and finance, must be mobilized. In such circumstances, agricultural products, especially grain products, should be cheap and available in sufficient quantities. The reasons for this are, first that grain provides the staple diet of the people. Cheap and sufficient grain supplies are thus needed to help maintain a low-wage labour force required by heavy industry. Second, cheap and available grain and other agricultural products are needed to acquire foreign currency in order to import more manufacturing facilities

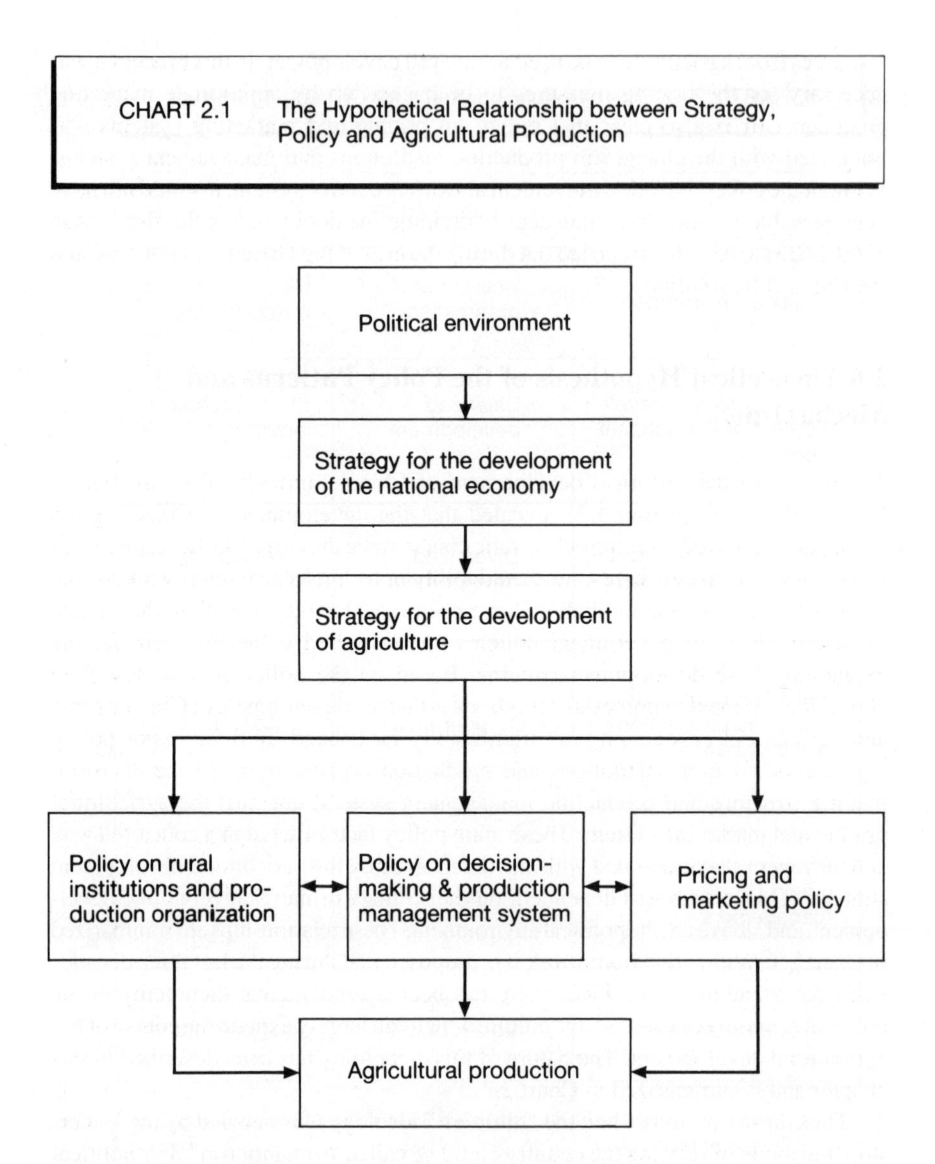

which are also required by heavy industry. Under this situation the strategy for the development of agriculture is a general squeeze but within this, priority is given to grain production.

To put these ideas and strategies into practice requires several policy ingredients. First, the agricultural production system has to be organized as a grassroots organization suitable for and responsible to the centralized command system. This necessitates collectivization. This, in turn, permits a direct control system to be followed in which producers are required to focus the limited resources on producing what is most needed by the government. At the same time, a low purchasing price set by the central government and the monopolized marketing

CHART 2.2 The two policy patterns for Chinese Agriculture

	Policy Pattern A	Policy Pattern B
Political environment	Realism and pragmatism	Ultra leftism and romanticism
Strategy for the development of the national economy	Balanced development	Stress on heavy industry
Strategy for the development of agriculture	Diversified development	Take grain as the link
Policy on rural institutions and production organization	Co-operatives	Collectives
Policy on decision-making and production management	Indirect control	Direct control
Policy on pricing and marketing system	Looser control and improved prices	Tight control and poor farm prices
Relevant eras	1952–1957 1961–1965 1978–1984	1958–1960 1966–1977

system can, in principle, lead to cheap and sufficient grain and other products to serve the aim of industrialization.

On the other hand, at times when the "ultra-left" ideology is criticized and a more pragmatic approach is pursued by the government in running the country (not infrequently this arises out of the bitter lessons of failure), a more balanced development strategy is pursued between agriculture, light industry and heavy industry. As a consequence, diversified development with a better use of natural and economic resources is pursued as the strategy for agricultural development.

Logically, this demands encouragement of farmers by means of a direct linkage between contribution and compensation, and the shift from collectivization to formation of co-operatives. By the same token, direct control has to be substituted by indirect control systems which give a greater degree of decision-making freedom to peasant farmers and basic productions units. This in turn allows for more locally responsive production management. Coincidentally, an increase in purchasing price and relaxation of the monopoly marketing system is required to make agricultural production profitable and attractive.

Two further matters become apparent from the theoretical framework proposed above. First, in hypothesizing the two alternative policy patterns, labels for dichotomies are made both in terms of key policy issues as well as the development strategies and political environment concerned (see Table 2.6). The justification for the labels for the three key policy elements was explained in sections 2.3, 2.4 and 2.5, they are reasonably descriptive and uncontroversial. Likewise, the labelling of the two development strategies, first for the development of the national economy, and second for agricultural development are based on the reality of periodical changes in emphasis. The most difficult issue is that of the proper labelling of the corresponding socio-political environment. As pointed out by White (1982) none of the dichotomies: ultra leftist/revisionist; revolutionary/pragmatic; radical/conservative; idealist/materialist; Maoist/Dengist are adequate to describe the complicated political development which occurred in China. Therefore, "all such dichotomies are convenient oversimplifications, which do not capture the complexity of political attitudes and alignments within the Chinese state" (White 1982, pp8–9). However, it is argued that, for the purposes of this study which is focussed on the grain economy, the dichotomies presented in Table 2.6 are a fruitful simplification which enable useful insights to be made. It is not the aim of this book to provide an exhaustive analysis of the political economy of rural China. The labels used for these dichotomies are thus to be understood as shorthand for the rather complicated issues summarized in the relevant sections of this chapter.

Second, following Lardy (1983), the policy review pursued and theoretical hypotheses proposed in this chapter concern the key policy measures implemented during different development periods rather than the personality behind the policy who might therefore deserve praise or blame for policy success or failure.

At the same time it is stressed that reality is always much more complicated than any kind of theoretical generalization or abstraction. The two alternative policy patterns proposed in this chapter are based on key policy issues identified and their periodic fluctuations. For example, the eras of the Great Leap Forward and the Cultural Revolution are treated together as Policy Pattern B in this analysis. This is done because they shared common features in the grain policies implemented. Thus policy went in the same direction and with similar consequences. On the other hand, these two periods were different in the extent of the policy changes and the seriousness of the damage created. Further details on these matters are discussed in Chapters Four and Five. In these chapters the two period oscillation idea is put to the test and their implications are teased out.

In summary, in this chapter, key policy issues concerning China's agricultural development have been identified in a step-by-step review. It has been suggested that during the last three decades since the founding of the PRC, although government policies changed from time to time and took different forms (and names) in different periods, two basic policy patterns can be identified. The policies pursued during the periods of 1952–57, 1961–65 and 1978–84 are classified as Policy Pattern A, and that for the periods of 1958–60 and 1966–77 as Policy Pattern B. Overall the length of time under each pattern has been roughly similar (18 years for Pattern A and 15 years for Pattern B).

Testing and validating these hypotheses will help clarify the erratic path of development of the Chinese grain economy during the last three decades. This will be achieved using quantitative analysis described in the next two chapters. It will become apparent that the complexities of the changes in institution, organization and price policy cannot be encompassed fully by relatively simple quantitative analysis. Thus Chapter Five provides a more detailed qualitative investigation of the main theoretical hypotheses. The result of this analysis is to provide answers to the first two questions raised in the Introduction.

Chapter Three

Analysing Chinese Agricultural Policy – Methodological Issues

3.1 Introduction

In stark contrast to the age of the agricultural sector itself, the history of the application of formal, quantitative, economic analysis to Chinese agriculture is very short indeed. The literature containing models of Chinese agriculture by Chinese scientists has only really developed since 1978. Apart from the disruption to research during the Cultural Revolution and the isolation of Chinese scientists, a substantial impediment to useful quantitative research has been the lack of a comprehensive, agreed, stable theory of producer and consumer behaviour under the Chinese centrally planned economic system. The models which have been built were mainly used for output projection and the selection of farming structure. Typical examples among them are the multi-objective mathematical programming model of the grain sector (Chen *et al.* 1983), the large scale linear programming model of grain and cash crops for the year 2000 (Grain and Cash Crops Development Research Group, Chinese Academy of Agricultural Science 1985 (hereafter CAAS 1985)) and the model of input-output analysis of Chinese agricultural sector of 1982 (Chen *et al.* 1985). These studies and the analysis described in this book all represent the application of Western analytical techniques to Chinese circumstances. The challenge is to discover which of these techniques may be usefully applied to a very different market system from that for which they were designed.

In the arsenal of sector modelling techniques developed in the West, three approaches are available for the research of policy issues: mathematical programming, simulation and econometric methods. Each technique has its own characteristics, and the choice depends on the objectives of the research, the nature of the system being analysed and the data available.

Programming models (see Hazell and Norton 1986) in their various forms can be used at the sector, region, nation, commodity, or resource levels. Large scale programming models are a powerful tool, especially for structural and policy impact analysis. However, the programming approach was not considered suitable to analyse the Chinese grain economy for several reasons. First, programming models, typically linear programming models, are often used to specify optimum or profit maximizing outputs at different levels of factor or product price. Profit maximization is the behavioural rule underlying producers' supply response and is explicitly built into the model specification. This is generally acceptable for commercial producers in developed countries, but it is less applicable to self-subsistence grain producers in China and other developing countries. Second, during the last few decades, production organization has been manipulated through government policy changing from individual producers (after the land reform) to co-operatives, then collectives and finally to the household responsibility system and integrated co-operation. It is, therefore, impossible to identify unique producers under this succession of changes. Moreover, since China is such a large country with such complicated farming systems, a very large number of representative farms or regions would be necessary to adequately cope with the whole sector. This would create an immense computational task. Third, the impacts of policy changes on agriculture invariably take many years to work through. As the prime objective of this study is analysis of the changes in policy on the grain sector during the last three decades, a static programming model using cross-sectional data is not suitable. Finally, manpower and financial constraints made it impossible to develop a large scale, dynamic programming approach.

Simulation modelling has been proposed as a useful tool in the agricultural sector in underdeveloped countries for evaluating the likely impacts of adjustments in agricultural structures and policies. However, this approach has not been greatly used in the modelling of the agriculture sector for more than ten years. The main reasons can be summarized as, first, the existence of a knowledge gap between economic and physical systems. A lack of knowledge of the effects of qualitative factors and an inability to quantify many of the inter-relationships of agricultural economic systems greatly hinders this approach. Second, specification of the appropriate functional relationships, and estimation of the relevant parameters are particularly difficult. In practice, these are considerably influenced by modellers' subjective judgements. Third, an interdisciplinary research team is required and cross-sectional and time-series data are needed for building and validating the simulation model. Thus the resource requirements are demanding.

The third broad approach, based on econometric estimation has been chosen for this project, mainly because of its positive nature, the availability of time-series data and its suitability for assessing the long-term impact of changes in policy on economic outcome.

3.2 The Broad Outlines of the Chosen Approach

The analytical technique used to test the hypotheses defined in the last chapter and thus to summarize the development of the grain economy is based on an aggregate econometric model. The justification for starting with an aggregate model lies with the ability to adapt it to Chinese circumstances and in the merit of the technique itself.

There are two major factors to be considered in adapting the approach to the Chinese situation. First, China is a socialist country with a planned economy. The historical development of the grain sector has been strictly controlled by government.

Changes in policy are therefore of paramount importance for the success or failure of the performance of the sector. However, as demonstrated by empirical research by Chinese agricultural economists, even under the situation of a direct planning system in which growing area is allocated from top to bottom, peasant farmers and grassroot production units still have scope to adjust production in several important ways according to their preferences (for details see section 5.1.2.2). For example, they can plant a larger or smaller growing area than they declare; they may differentiate between crops in allocating different qualities of land; they can interplant one crop with another in the same piece of land; change the allocation of physical and labour inputs for different cropping activities (this involves discriminating in quantity, quality and timeliness); and finally they may choose between new technologies for reasons of profitability or risk. In short, the combination of the nature of the government policy and producers' room for manoeuvre under certain circumstances may be manifested in a way which can usefully be analysed at the sector level.

Second, as described in Chapter Two, rural institutions and production organization are products of government policy and as policy has changed, so also has the basic production unit. For aggregate econometric modelling with time-series data, there is no requirement to specify unique producers at the grassroots level.

Thus, the reasons aggregate time-series analysis is the preferred analytical approach may be summarized as: (i) It operates directly upon the aggregate data which are the objects of interest for projection purposes. (ii) It is the simplest of the procedures in terms of estimation methods and data requirements. (iii) It is a technique which has shown itself capable of "generating acceptable and useful results" (Colman 1983).

Modelling the Chinese grain economy in this study includes the specification and estimation of supply response, consumption and trade functions. The supply response analysis has been done both at the aggregate sector level and regionally. The regional supply response analysis is intended to provide more disaggregated results to help validate the aggregate model. Together, these provide the answers to questions (a) and (b) posed in the Introduction. Following the supply response analysis, attention turns to specification and estimation of consumption and then trade. The major hypothesis relating to the consumption function is that consumption of food grain is not only dependent on consumers' income level and selling price, but also, and perhaps more importantly, on the production level achieved and the growth rate of population. Supply and demand then work together to determine grain trade. For these reasons, a recursive econometric model system was built, tested and discussed to throw light on the third question raised in the Introduction.

Before summarizing and presenting the results of the empirical analysis it is worth airing two general issues concerning a fundamental difference in philosophical approach. First, Western econometric models are based on a coherent theory of profit maximizing behaviour of producers and utility maximizing behaviour of consumers under the operation of the market mechanism which may or may not be interfered with by government. This dictates the variables to be included in model specification and the interpretation of quantitative evidence from the estimated models. However, in applying such techniques to Chinese agriculture, these assumptions are clearly inappropriate. The rôle of the quantitative techniques is thus to clarify the essential factors which influence Chinese grain supply, consumption and trade and in turn to test the validity of the broad theoretical hypotheses proposed. Therefore, the approach pursued in this book is more pragmatic and simulative in nature than would be the case in the Western literature. Second, it should be remembered that early quantitative analysis in the West started in a similar rather pragmatic way as adopted here. The emergence of the full, modern theory of production and consumption in the West which nowadays makes great use of duality theory has been a result of the combination of empirical econometric analysis and abstract theoretical analysis. This process is in its infancy as far as the Chinese command economy is concerned. This book may be regarded as a small step in the same direction for China.

3.3 Supply Response Analysis in General

Supply response analysis has been a field of interest to Western agricultural economists for more than half a century, and is currently being pursued with greater vigour than ever before. The objectives of research on changes in agricultural supply, according to Nerlove and Bachman (1960, p532), are to improve, understanding of the mechanism of supply response, ability to forecast supply changes, and competence to prescribe solutions to problems related to agricultural supply. Staniforth and Diesslin (1961, pp293–294) spelled out objectives, in somewhat more detail, first to provide guidance for general policy formulation. This primarily involves drawing conclusions of probable aggregate changes in output and direction of necessary changes in output, in inputs and firm structure. Second, they suggested it could indicate how much relevant variables should be manipulated to effect specific changes. The large literature on supply response analysis has been reviewed by Nerlove and Bachman (1960), Heady *et al.* (1961), Cowling and Gardner (1963), FAO (1971) and Colman (1978, 1983).

In the absence of a fully specified theoretical model of individual producer behaviour and an appropriate aggregation procedure to translate this to the sector level, the supply model is specified in a rather *ad hoc* way directly at the aggregate level. The single equation, econometric method is then employed in the estimation of a supply response function first for the grain sector as a whole and then for the individual crops of rice, wheat, soybean and other grain. In addition regional analysis for the grain growing areas of Shandong and Fujian Provinces is performed to help validate the aggregate model.

The term "agricultural supply response analysis" is defined by Colman (1983, p201) as, "the art of estimating the quantitative supply response of agricultural commodities to changes in product prices, input prices and other relevant measurable aspects of the changing environment for agricultural production". Based on neo-classical economic theory, single equation supply response was typically specified to explain the response of area planted or harvested to economic signals

(Nerlove 1956, 1958, 1979; Colman 1972; Tomek 1972; Fisher and Tanner 1978; Sanderson *et al.* 1980).

For example, Colman (1972, pp49–51) specified equations for the supply of cereals in Britain as:

$$Ait = f[P'it, P'jt, Yit, Yjt, P'kt, Ykt, Pnt, Ut] \quad (3.1)$$

where A = area harvested
Y = yield per hectare
P = price, including subsidy, per unit of output
P = price per unit
U = disturbance term
i = wheat, barley, oats or mixed corn
j = the set of cereals excluding the endogenous one
k = the set of alternative agricultural products
n = the set of agricultural inputs

Since the number of explanatory variables is too large, the function was simplified as,

$$Ait = f\,[Rit, Rjt, Rkt, Cit, Cjt, Ckt, Ut] \quad (3.2)$$
$$\text{where } Ri = Yi.p'i,\ Rj = Yj.p'j,\ Rk = Yk.P'k$$

Because of the unavailability of cost data and the consideration of capital fixity in agriculture, a partial adjustment mechanism was introduced, and the general area response function was rewritten as,

$$Ait = f\,[Ait\text{-}1, Rit, Rjt, Rkt, Ut] \quad (3.3)$$

After taking into account the influence on decision-making of the time gap between crop cultivation and harvesting, a one period lag was added to the variables on the right-hand side of the equation (3.3). The area response equation thus becomes,

$$Ait = f\,[Ait\text{-}1, Rit\text{-}1, Rjt\text{-}1, Rkt\text{-}1, Xt\text{-}1, Ut] \quad (3.4)$$

where X refers to other factors affecting ploughing and planting the crop.

Three justifications are given for this specification of the supply model:

First, it is based on the theory of the neo-classical firm in a market economy. Profit maximization is assumed to be the aim pursued by commercial producers. If there are large numbers of producers, none producing a significant share of the industry output, and if there is relative freedom of entry to and exit from the industry, the market structure is labelled perfectly competitive. This in turn means that individual producers are price takers in selling products and buying inputs. Producers therefore have only to make decisions about the composition and volume of output and the quantities of inputs to be employed. In turn, this suggests that the market price of the goods to be produced, the prices of competitive crops and of agricultural inputs are the three major signals for producers in their decisions on "what", "how" and "how much" to produce.

Second, for individual producers given technological conditions, the foremost production decision is the allocation of land between competing enterprises.

Third, this general specification explicitly allows for the uncertain nature of crop production. The uncontrollable influences of weather and disease create yield variability. This, combined with the variability of consumption and the interaction between markets for different crops results in price instability. Government actions often create a third tier of uncertainty. For these reasons price variables are specified as expected prices. In turn, there are numerous formulations of price expectations, the simplest of which is the naive model which suggests that next year's expected price is this year's actual price.

3.4 Supply Response Analysis for China

The conventional specification of a supply response function is rooted in neo-classical economic theory, that is, the theory of profit maximizing behaviour of agricultural producers under the operation of the market mechanism. Because of the fundamental institutional and managerial differences between China and market economies of the West, it is necessary to modify the specifications of the supply function to adapt to China's circumstances and meet the particular objectives of this analysis. Five features are identified which define the special characteristics of the Chinese supply model. These are: a genuinely sectoral definition of supply, a focus on output not area, the workings of the price system, the importance of self-subsistence and the formation of expectations.

A truly sectoral approach As was discussed in Chapter Two, government policy has not only played a key and direct rôle in controlling the production and consumption of agricultural produce as well as industrial inputs, both in quantity and price, but it also determined production organization and decision-making in agricultural production. Thus government policy has attempted directly to determine the level and composition of agricultural production. In these circumstances it seems reasonable to conceive of a supply function which represents a sectoral response *per se* and is not a simple aggregation of individual producer response. A further, practical, supporting argument for this interpretation is that with the series of changes in production organization the very definition of producer decision makers change from time to time from teams to co-operatives to collectives to individual households. This makes it an impossible task to conduct a supply response analysis based on individual producers for the last three decades.

Output not area It makes more sense to define output as the endogenous variable rather than growing area. China has had low per capita grain production for the last three decades. In 1953 when the first Five-Year Plan was initiated, the per capita grain availability of the nation was 284 kg. Twenty-five years later, in 1978, it was 317 kg. Despite the nation's total grain output of 387 million tonnes in 1983 which made China the largest grain producer in the world, per capita grain production of 387 kg was still below the world average of 400 kg. Thus China has continually struggled to increase in grain production in order both to speed up the industrialization programme and to provide basic subsistence for the huge and growing population. To do this, both grain growing area and measures to improve crop yields were identified as principal targets for government action. With the operations of a planned economic system, annual production plans were drawn by the central government with specified grain growing area and measures for yield improvement. These plans were passed down step-by-step. When the "direct control" system was pursued during the periods of Great Leap Forward and Cultural Revolution, these targets were to be implemented directly by the basic production units (see section 2.4.2). It is therefore appropriate to take output as the dependent variable in modelling grain supplies rather than growing area. In addition, from the sector supply response analysis point of view, growing area could be misleading if used as the endogenous variable because of the declining trend in growing area and changes in the intensity of land use through multiple cropping.

The price mechanism This is fundamentally different to that of a market economy. First, as reviewed in Chapter Two, the pricing and marketing system was directly controlled by the central government. With the operation of the compulsory purchasing system, prices for grain and other main agricultural products were fixed and a compulsory quota system was imposed. The prices for major production inputs, such as chemical fertilizers, are also determined by the government with the annual level of supply specified. Pricing policy was integrated with the other elements of social control of agriculture, that is, the policies on rural institutions and production organization and the policies and mechanisms for decision-making and production management. The combination of these elements together formed unique policy patterns under which the agricultural sector progressed.

Within this complex package, the adjustment of price ratios between grain and its major competing crops was used by the government as a method of pursuing alternative development strategies during different periods. For example, when "taking grain as the link" was pursued as the strategy for the development of the agricultural sector, the price ratio between grain and some other agricultural products (typically the cotton crop) was adjusted in favour of grain production. At the same time the policy pattern pursued was a general squeeze on agriculture as shown by poor overall farm prices. However, within this squeeze was an emphasis on grain production (see section 2.6). In contrast, when "diversified development" was encouraged, a significant improvement in purchasing prices for grain and other agricultural produce was offered with more favourable conditions for cash crop production. These price policies show up as changes in the price ratio between grain and cotton. This will be discussed in detail in the next chapter.

Subsistence Around three quarters of the grain produced in China is for peasant farmers' self-subsistence rather than commodity production. Self-subsistence is therefore a key factor influencing peasant producers' supply response behaviour in addition to considerations of profitability. The self-subsistence orientation is due not only to the weakness in social infrastructure but also significantly compounded by the imposition of a compulsory purchasing system and the centralized distribution system (for a review of the grain distribution systems, see section 6.2). In these circumstances, most basic production units and peasant households obtain their subsistence grain by producing it themselves. This is especially so during the periods when the Policy Pattern B was pursued.

Expectations On the face of it, because the price set by the government for official purchasing was predetermined and announced before each annual production cycle, there seemed to be no need to introduce a price expectation mechanism in the specification of the supply function. However, it must be acknowledged that promised prices and quantities were not always fulfilled. Thus farmers had to formulate their expectations of the relationship between promised and actual price. The use of simple lags did not seem an adequate description of such expectation formation, thus lagged variables were not used.

All the main features explored above indicate that the supply response mechanism for China's grain sector is quite different from that under a free market mechanism. It should also be clear that in these circumstances, the scope for the nation's grain producers in responding to price signals is significantly constrained.

The discussion above might be thought to lead to the conclusion that under a command economic system, there would be little room for grain producers to respond to the price changes. This is not so. When the "tight control, poor farm prices" pattern was pursued, agricultural producers' initiative in production was seriously dampened, consequently, production of grain and other agricultural products either crashed or stagnated. In contrast, when the policy pattern involved "looser control, and improved prices", producers' enthusiasm for agricultural production was effectively mobilized, and therefore significant production growth both in terms of grain as well as other major agricultural products was registered. At the grassroots level, although the whole range of government intervention did create serious constraints on individual producer's price response, peasant farmers did have means available to them to fight to protect their own interests. Since the government policy for China's agricultural development oscillated periodically between Policy Patterns A and B during the last three decades, the scope for producers to respond to price signals changed considerably. Generally speaking, peasant farmers enjoyed a greater degree of freedom in response to price changes when Policy Pattern A was adopted. These ideas are put to the test in the next chapter.

Part II

Empirical Analysis of Historical Development

Chapter Four

Grain Supply Response – A Statistical Analysis

Having set out the philosophy and general approach in Chapter Three, this chapter contains the results of applying simple quantitative techniques to aggregate data for the Chinese grain sector. It is first necessary to discuss the variables used in the analysis and the main features displayed by these variables over the period 1952 to 1984. This is followed by discussion of the results for the grain sector as a whole, then rice, wheat, soybeans and other grains. The final section summarizes two regional analyses which were undertaken to validate the approach at a lower level of aggregation.

4.1 The General Supply Model and the Variables

The arguments of Chapter Three concluded that an appropriate general theoretical model for the supply response of China's grain sector could be specified as a single equation for each grain type in which output depends on:

The purchasing prices of all relevant grains offered by the official marketing agency (denoted P),

The selling prices of all relevant inputs determined by the official marketing agency (P1),

The grain growing area (A),

A time trend representing technological improvement (T),

An indicator variable for the changes in natural conditions (W),

Intercept dummy variables for different time periods (D).

The dependent variable Grain in China includes rice, wheat, coarse grains, tubers and legumes. Amongst these, paddy rice, wheat, corn, soybeans and sweet potato are the most important grain crops. In China's official statistics, grain output is measured in unmilled grain. The conversion ratio from tubers to grain equivalent was 4:1 before 1963 and 5:1 since 1964. The analysis was conducted for grain as a whole, and also disaggregated into rice, wheat, soybeans and other grain crops.

This latter category included corn and sweet potato together as data was not available to separate them for the late 1960s.

Grain output is the product of growing area and yield. For some time now, growth in yield has frequently been taken as a prime indicator of the level of development and the performance of party and governmental organs. This was especially so when "taking grain as the link" was the policy for agriculture. As a consequence, a larger growing area than the planned target area from the upper level was often requested by local officers in order to increase output. This meant the official figures for grain area were often in error. However, it is more difficult for local government to manipulate the figures for grain output. This is because output is the main evidence used in deciding the amount of grain to be purchased and exported from the region at annual grain procurement meetings. As a result, data for the grain output is comparatively more accurate than growing area and is therefore used as the dependent variable (see the discussion in section 1.6 which assessed some of the problems of data quality).

Turning to the independent variables, the output response model is specified as a function of three major components, government policies, technological improvement and the natural environment. Government policies are represented by prices, price ratios and dummy variables.

Grain purchasing price Data on the grain purchasing price which was used for the sector supply analysis refers to the weighted average of quota price, above-quota premium and negotiated price, all offered by the official marketing system (see Appendix Table 1.3). Since the pricing policy for official purchasing changed from time to time, there are periods for which there was no quota premium and negotiated price. Therefore, it is impossible to differentiate the effects of these various types of price in the aggregate time-series analysis. Also, as the official grain purchasing made up an overwhelming part of the total grain traded for the whole period of the last three decades and reliable, comprehensive time-series data for the price of free market trade is not available, these prices were not taken into consideration.

The main price signal for producers' supply response is hypothesized to be the grain purchasing price. As discussed above, unlike commercial producers under a market economy, the grain market in China was totally or predominantly controlled by the official marketing system. The purchasing price was fixed and a compulsory quota system imposed by the government. Furthermore, supply and demand were separated and the government did not only work as a unique purchaser but also as a monopolist supplier of grain products. Because the supply response mechanism for China's grain sector is fundamentally different from that of a market economy, this makes it important to emphasize several distinctions. First, that producers do not have many degrees of freedom to respond to price signals mainly because of the isolation of supply and demand. This is particularly so when the direct control system was imposed and the free market prohibited. Second, by the same token, the purchasing price is regarded as the key signal of profitability. Clearly it should not be amalgamated with the other two factors (the prices of competitive crops and major inputs) neither is it wise to simplify it as two price ratios which is quite often the approach taken by empirical research (see further discussions later). Third, there is no need to introduce a price expectation mechanism in the model for the reasons explained above. Finally, since the agricultural pricing and marketing system was one of the major policy ingredients of the policy patterns under which China's grain and agricultural sector developed, the changes in grain purchasing price was generally consistent with the shifts in policy patterns. When Policy Pattern A was pursued, a significant increase in grain purchasing price was offered like that for other major agricultural products. In contrast, when Policy Pattern B was adopted, a general stagnation in purchasing price was recorded (see section 2.5). Therefore, it is expected that there is a positive relationship between grain output and the grain purchasing price.

Prices for competitive crops and major inputs For Chinese grain production as a whole, the main competitive crop is cotton, and the major industrial input is chemical fertilizer. While the markets for grain and other major agricultural products were directly controlled, the price ratios between grain and cotton products and between grain and fertilizer were also manoeuvred by the government as a means of pursuing alternative development strategies. For example, when the national economic development strategy was to emphasize heavy industry and when the strategy for agricultural development was to "take the grain as the link", agricultural collectivization was pursued with a direct production control system and pricing policy inflicted a hard squeeze on agriculture whilst, within this, favouring grain production. Therefore, the purchasing price ratio between grain and cotton, the main competitive crop in China's circumstances, was manipulated against cotton production and in favour of grain. On the other hand, when balanced development was stressed and diversified development was the strategy for the agricultural sector, the price ratio between grain and cotton was adjusted in favour of cotton production. Significant price improvements were then given both for grain and other major agricultural products. Thus, the general situation during the last three decades was that when Policy Pattern A was pursued, the price ratio between grain and cotton fell and was generally lower to encourage more diversified development in agriculture. In contrast, when Policy Pattern B was adopted, the price ratio of grain and cotton rose or was generally higher in pursuit of the strategy of "taking grain as the link". This pattern is shown in Tables 4.1 to 4.3 and Charts 4.1 to 4.4. Since rice and wheat are the two most important grain crops, the trend is more clearly shown in Charts 4.2 and 4.3. Coincidental with the changes in price ratio of grain and cotton, the price ratio between grain and fertilizer was also adjusted by the central administrative system, that is, in favour of grain production when Policy Pattern B was pursued and less so when the alternative policy pattern was implemented. Since fertilizer supply directly manoeuvred by the central government made up only part of the total available supplies, the movement of the price ratio between grain and fertilizer does not reflect the hypotheses pattern as clearly as that shown by the price ratio of grain and cotton.

In addition to the independent policy variables of the grain purchasing price and growing area, two price ratios are employed to simulate the shifts in policy patterns.

Table 4.1 Trends in the output of grain and the price ratio of grain to cotton

Year	Changes in grain output		Changes in price ratio of grain/cotton	
	Grain output (10,000 tonnes)	Increase in %	Price ratio of grain/ cotton in %	Increase in %
1953	16,683		90.9	
1957	19,505	+16.9	90.2	–0.8
1961	14,750	–24.4	117.0	+29.7
1965	19,453	+31.9	112.4	–3.9
1977	28,273	+45.3	123.4	+9.8
1984	40,731	+44.1	115.6	–6.3

Source: Derived from Appendix Table 1.1 and 1.3

Table 4.2 Trends in the output of rice and the price ratio of rice to cotton

Year	Changes in rice output		Changes in price ratio of rice/cotton	
	Rice output (10,000 tonnes)	Increase in %	Price ratio of rice/ cotton in %	Increase in %
1953	7,127		69.9	
1957	8,678	+21.8	68.8	–1.6
1961	5,364	–38.2	90.7	+32.2
1965	8,772	+63.5	83.0	–8.5
1977	12,857	+46.6	94.6	+14.0
1984	17,826	+38.6	70.5	–25.5

Source: Derived from Appendix Table 1.1 and 1.3

This somewhat unconventional approach is based on two major considerations. First, as mentioned above, the price ratios were manipulated as an instrument in pursuit of development strategies. Therefore the changes in price ratios (typically that of grain and cotton) coincided with the shifts in policy patterns. Second, policy on agricultural pricing and the marketing system was an organic component of the policy patterns and it was integrated with the corresponding policies on rural institutions and the production control system. In these circumstances, as long as there existed a price response effect, it would be the purchasing prices of the

Table 4.3 Trends in the output of wheat and the price ratio of wheat to cotton

Year	Changes in wheat output		Changes in price ratio of wheat/cotton	
	Wheat output (10,000 tonnes)	Increase in %	Price ratio of wheat/ cotton in %	Increase in %
1953	1,828		108.8	
1957	2,364	+29.3	99.4	-8.6
1961	1,425	–39.7	126.0	+26.8
1965	2,522	+77.0	108.4	–14.0
1977	4,108	+62.9	129.1	+19.1
1984	8,782	+113.8	96.1	–25.6

Source: Derived from Appendix Table 1.1 and 1.3

individual crops which would be the most important matter, e.g. the shifts between "poor farm prices" and "improved prices" are the most critical signals to which the sector responds.

While a positive relationship is expected between the grain purchasing price and grain output, a negative coefficient is expected for each of the two price ratios specified. The explanation for this seemingly contradictory phenomenon is that it is the shifts between "poor farm prices" and "improved prices" which is most critical in terms of price response effect in the grain sector. For example, during the eras of the Great Leap Forward and the Cultural Revolution in which Policy Pattern B was pursued, the nominal purchasing prices for grain and other major agricultural products were stagnant for most of the time and, allowing for inflation, the real prices for some actually declined. Such "poor farm prices" were geared to a strict market control and compounded by the implementation of agricultural collectivization and a direct production control system. The fundamental socio-economic environment thus discouraged the development of grain as well as agricultural production more generally. In this case, the generally unfavourable conditions for grain production could not be significantly changed by means of adjustment in price ratios between grain and other major agricultural products. Therefore, peasant farmer's initiative in the development of grain production was seriously dampened. Consequently, grain production of the nation stagnated during the Cultural Revolution (in terms of per capita production) and crashed during the era of the Great Leap Forward. So this relatively high, or rising, ratio of grain to cotton prices in these periods coincided with relatively low or falling grain output. The converse applied in the periods during which Policy Pattern A applied. That is a lower or falling ratio of grain and cotton prices which coincided with a significant increase in grain output. Hence the coefficient of the price ratio variable is expected to be negative.

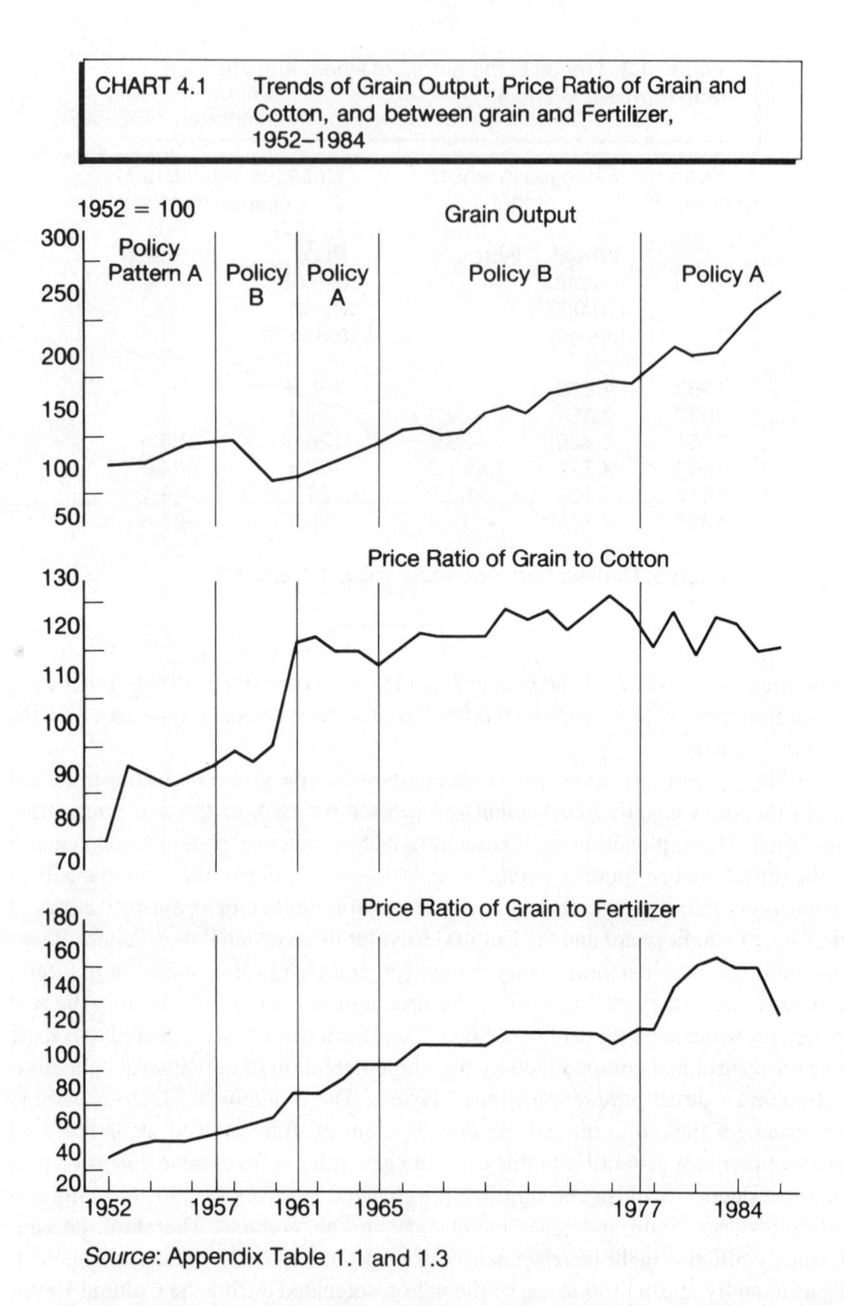

CHART 4.1 Trends of Grain Output, Price Ratio of Grain and Cotton, and between grain and Fertilizer, 1952–1984

Source: Appendix Table 1.1 and 1.3

Grain growing area The grain growing area is considered to be another important factor partly because it was the core among grain production planning targets and the key figure from the government point of view for ensuring the achievement of expected output level and the realization of the planned purchasing target. More importantly, due to the limitation of arable land, the low level of per capita grain production and the dispersion of the market, a minimum grain growing area to serve

CHART 4.2 Trends of Rice Output, Price Ratio of Rice and Cotton, and between Rice and Fertilizer, 1952–1984

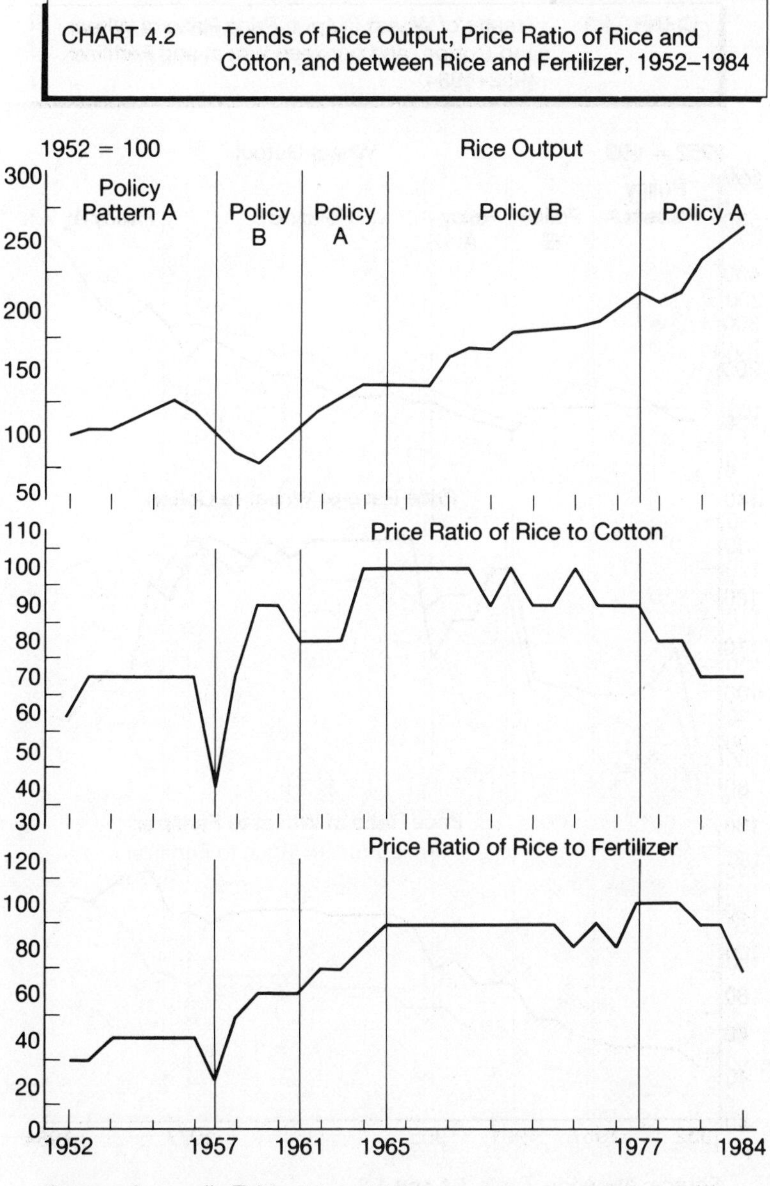

Source: Appendix Table 1.1 and 1.3

self-subsistence requirements is also of concern to the basic production units and peasant farmers.

There are two main issues inherent in the official statistics on grain growing area. One is that when the strategy of "taking grain as the link" was pursued, a greater growing area than the planned target set by the upper level was often requested by the local officers in order to declare a higher yield achievement. There is therefore

CHART 4.3 Trends of Wheat Output, Price Ratio of Wheat and Cotton, and between Wheat and Fertilizer, 1952–1984

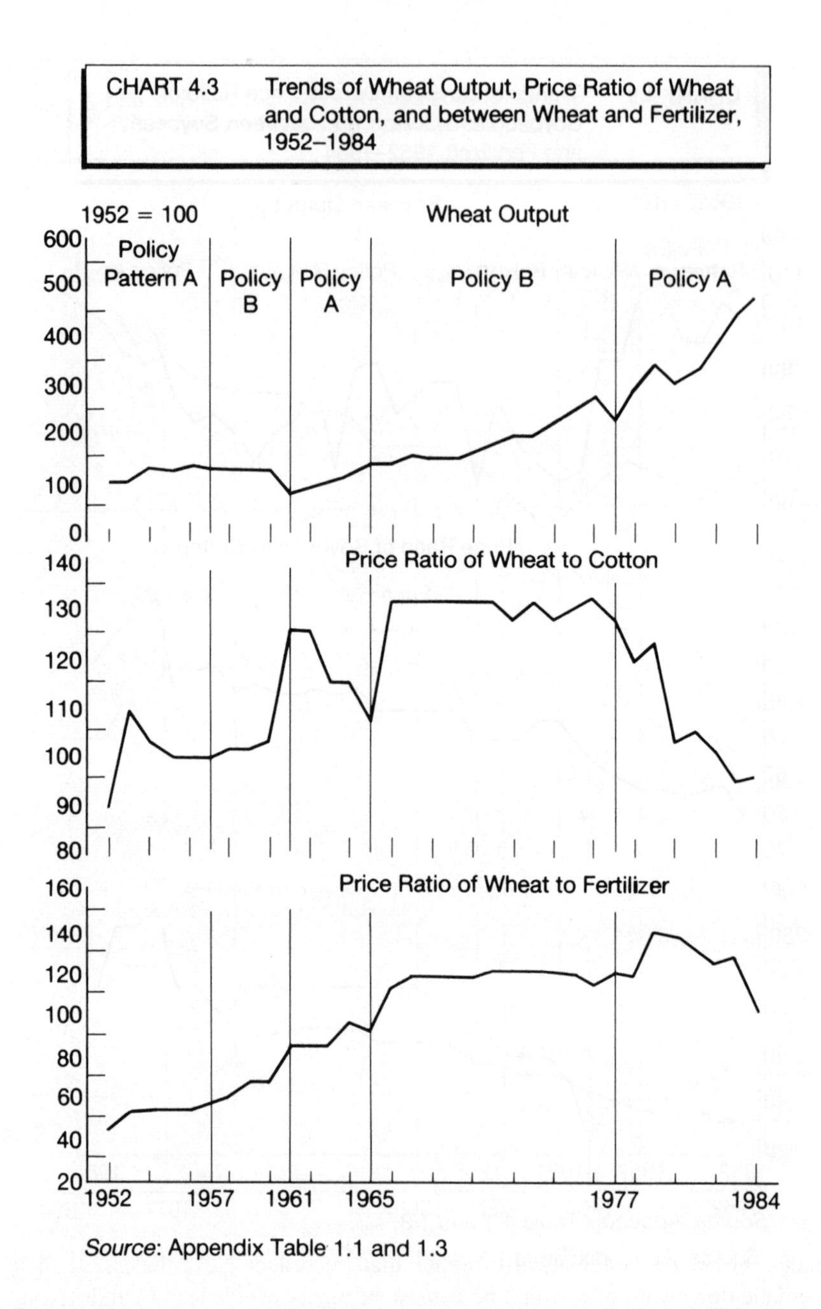

Source: Appendix Table 1.1 and 1.3

an expectation that the growing area recorded was lower than actual during the eras of the Great Leap Forward and the Cultural Revolution. The other concerns the general accuracy of the official statistics for arable land. According to the recent report presented by the National Agricultural Zoning Committee of China (NAZC), the actual cultivated land of the nation is about 30% higher than the official figure (*NAZC 1985*, p52). If this is the case, the real grain growing area during the last three

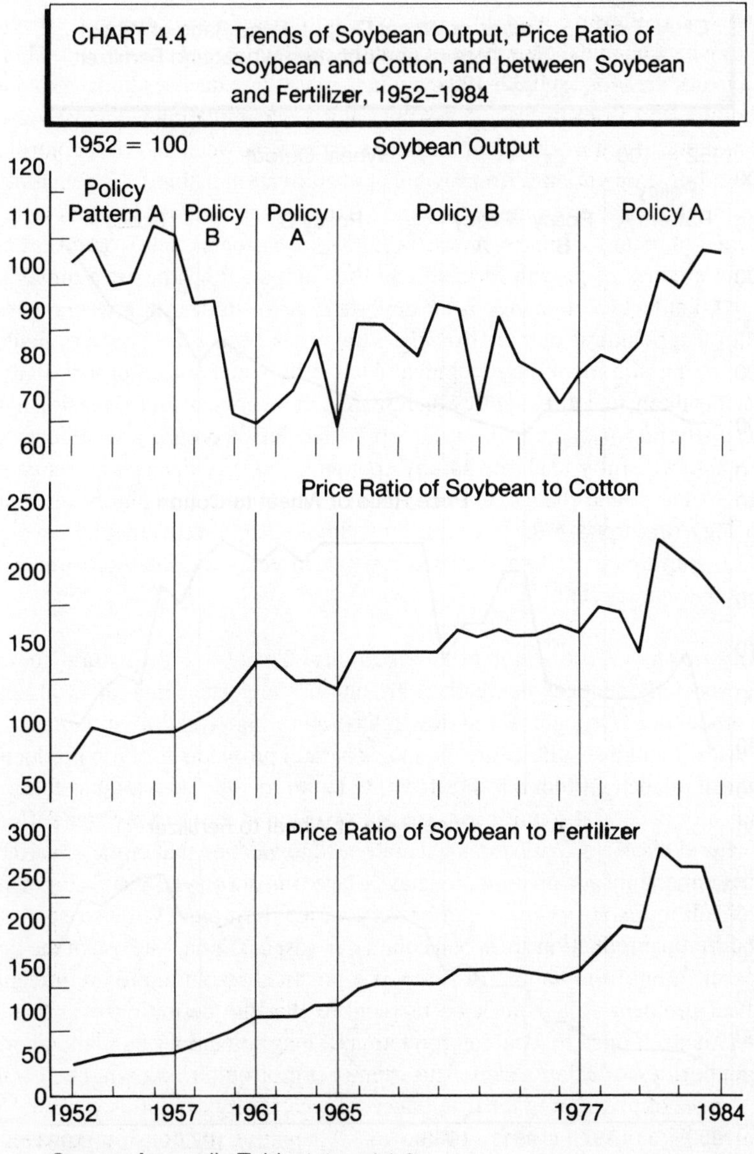

CHART 4.4 Trends of Soybean Output, Price Ratio of Soybean and Cotton, and between Soybean and Fertilizer, 1952–1984

Source: Appendix Table 1.1 and 1.3

decades would be systematically higher than recorded. Nevertheless, such a systematic downward bias would be consistent for the whole historical development period and simply means the actual yields were considerably lower than the official figure. Because there is no other time-series data available, the official figures on grain growing area are used in this analysis.

Technological improvement Due to the high population/land ratio, intensive farming is one of the main characteristics of Chinese agricultural production.

Technological improvement in grain production has been stressed especially changes which enable the improvement of the productivity of arable land. This has been a major area for capital investment in the last three decades both from public expenditure and from private, grassroot capital accumulation. Increased use of modern inputs, such as chemical fertilizers, irrigation, weed and pest control and high-yielding crop varieties; improvement in cultivation methods, for example by increasing the multi-cropping index; and extension and training services were the main areas of focus for improvement. Technological progress has worked as one of the main engines of growth in yield and thus output. It has become universally accepted that technical change is an important factor in determining the pace of growth of agricultural output. However, there has been little progress made in characterizing and modelling technical change by means of empirical analysis. Whilst it is likely that direct policy on research, development and extension has an impact on the rate of emergence and adoption of technical progress, the effects have not been successfully analysed in any economic system. For these reasons, the influence of technical change on grain production is handled in a purely descriptive way using a time trend as the indicator for technological improvement. Like that of grain growing area, a positive relationship between technological improvement and grain output is expected.

Indicator for changes in natural conditions Changes in the natural environment, especially changes in weather, are another important ingredient affecting grain production. Drought, flood, low temperature injury and frost damage, and plant disease and insect pests are the major natural problems for crop production. The extent of such problems varies from region to region. The variable chosen as the indicator of natural conditions was the proportion of the growing area suffering from natural disasters. Whilst it is generally acknowledged that climatic variables such as temperature and precipitation have a large impact on yields and output, they are difficult to quantify. These can be very varied during the same season in the many different production areas of a country as vast as China. The use of the index of area suffering from natural disasters is a practical resolution to an intractable statistical problem, but it must be recognized that the definition of a "natural disaster" is itself open to local interpretation. It may sometimes be in the interests of the authorities to declare a natural disaster when crop output is down. There is thus some danger of overstating the true effects of weather by using this variable. The magnitude of any such bias is not known. A negative relationship between the weather indicator and grain output is expected.

Dummy variables to indicate policy pattern change It is suggested that the independent variables discussed above are the most important factors influencing the historical development of China's grain production. Therefore, the natural, technological and socio-economic environment of the nation's grain supply have been accurately modelled. However, five separate development phases have been identified for the whole period of 1952–84. These are 1952–57, 1958–60, 1961–65, 1966–77 and 1978–84 (discussion of the justification for this division is contained in Chapter Two). An equation specified without dummy variables imposes the restriction that there is no significant difference between the five individual periods. Nevertheless, it is possible that there are impacts of the policies pursued in these periods which are not completely captured by the explanatory variables discussed so far. The truth of this can best be tested by specifying and estimating alternative less restricted equations. This can be done by introducing dummy variables for individual periods and then testing the validity of the restrictions imposed. In doing this the specifications of the dummy variables does not exactly coincide with the periods defined in Chapter Two because allowance has to be made for time lags in the impact of policy change.

Another issue which requires discussion in model specification is whether or not to deflate the economic variables and by what deflator. In supply and demand analysis, deflating usually means dividing nominal prices and incomes by a price index to get real prices and incomes. In practice, there is no unique rule to follow in making such a decision (see Tomek and Robinson 1981, pp319–322). Dealing with the Chinese situation a rather flexible attitude is taken and the decision made according to three major considerations, (1) the availability of data (for example, a nominal data series is used when it is the only statistic available for a particular variable i.e. there is no suitable price index for deflating). (2) In certain circumstances it is not necessary to deflate, for example, it is believed that the price ratio between animal products and feed grains is an important variable for livestock producers, the ratio instead of individual variables would be taken in the feed demand equation (see section 6.3.3). Clearly, in this case there is no need for any other deflation of the variables. (3) Prior knowledge of China's economy and statistics.

There are three relevant facts concerning inflation. One is the official denial of its existence in a socialist economy. For most of the time before 1978, people were encouraged by professionals and politicians to believe that inflation is a disease of the market economy in capitalist societies. Under a socialist planned economy, there is no room for inflation. The second is the fact that statistical series on inflation were not published for most of the 1960s and 1970s. So even if there were suspicions about inflation, there is little objective, comprehensive statistical evidence to support it. The third concerns the measurement of inflation. There are substantial grounds to doubt the accuracy; it is expected that there is a downward bias in the price index compiled. There are various reasons for this, for example, the prevailing hostile official attitude toward inflation; the fact of the inclusion in the index only of staple goods which are directly controlled by the administrative system, and the political concern about the relations between the state and the peasant farmers, the state and local governments, and the state and the public in general. All these contribute to a downward bias in relevant statistics published. In terms of producer and consumer's response behaviour to the change in price, an illusion effect which means supply or demand of goods changes in response to nominal price rather than real price could be, and indeed was, created.

With these considerations in mind, a good deal of judgment was exercised in the choice of nominal or real prices. Nominal data series are usually chosen for the reasons discussed. Where an appropriate price index was available both deflated and undeflated prices were tried in model estimation. Supply response equations with deflated grain purchasing price will be presented in the following section to compare with the corresponding equations with nominal prices.

Before considering the estimation of the supply relationships and the results, it must be stressed that the purpose of the grain sector supply response analysis discussed here is to identify the supply response mechanism under Chinese circumstances and to test the theoretical hypothesis concerning the policy patterns proposed in Chapter Two. In doing so, the philosophical approach adopted in this analysis is quite different from that for a market economy. This is the fundamental distinction between this research and correspondent empirical work on agricultural supply in the West. Bearing this in mind, individual results from the estimation, such as parameters estimated and elasticities derived must be interpreted with caution.

4.2 Model Estimation

As pointed out in the last chapter, the model estimation for grain supply response involves both sector and regional aspects. The sectoral analysis includes the grain sector as a whole and is then disaggregated into rice, wheat, soybean and other grain crops. For the regional analysis, two grain growing regions, one from north China (Pindu County of Shandong Province) and the other from the south (Jianou County of Fujian Province) were analysed. The statistical technique employed was Ordinary Least Square (OLS) and this was performed using the Times Series Processor (TSP) computing package.

Estimation was performed in both original variables and in first differences with both linear and log-linear functional forms. The results were assessed and tested from a statistical point of view and according to their conformity with the hypothesized behaviour of Chinese grain producers. The estimations with first differences were generally unsuccessful and the equations with linear functional form always outperformed those in log-linear form. Thus unless a special note is made, the equations presented are for the original variables in linear form only.

A number of considerations were relevant in choosing the appropriate price ratios for individual grain crops. First, because of the geographical location of the major grain crops and the rotational system existing in China, the internal substitutions between rice and wheat and other grain crops are not significant. On the other hand, soybeans is a small grain crop with a commercially orientated production motivation and similar regional and rotational distribution pattern as wheat and corn crops, thus competition between soybean and wheat and corn crops is considered in the model specification and estimation of the soybean output response equations. Finally, slope dummy variables were incorporated to test if there are also differences in estimated coefficients of independent variables between different development periods. However, as the results showed no significant coefficients, they are not presented or discussed.

4.2.1 The data

Research on Chinese agricultural development has always been hampered by data problems. This is particularly so for research conducted outside the country. The data used for the sectoral analysis was all published by the Chinese Statistical Authority (the SSB) and is contained in the 1983 and 1985 editions of the *Chinese Statistical Yearbook (TJNJ)* and *Statistics of Agricultural Economy 1949–1983 (Nongye Jingji Ziliao 1949–1983* (hereafter *NJZL 1949–1983*)), compiled by the Chinese Ministry of Agriculture, Animal Husbandry and Fisheries (hereafter the MAAF). Data for two grain growing regions are provided by the corresponding County Planning Commissions. The time-series data used in modelling are for 1952–84 and are presented in Appendix One.

4.2.2 Sector output response estimation

4.2.2.1 The grain sector as a whole

Based on the general formulation described above, the empirical model for estimation specified that the output of grain is explained by:

Grain purchasing price
Price ratio of purchasing price of grain over cotton
Price ratio of purchasing price of grain over selling price of fertilizer
Grain growing area
Crop growing area which suffered from natural calamities resulting in 30% or more reduction in yield
Time trend variable for technological improvement
Dummy variables for different policy eras, $D_1 = 1$ for 1959–61, $D_2 = 1$ for 1962–65, $D_3 = 1$ for 1966–77, $D_4 = 1$ for 1978–84.

Results from the model estimation are presented in Table 4.4, in which equations 4.1–4.3 are those of restricted equations and equations 4.4–4.7 the unrestricted equations corresponding to equation 4.3. Bearing in mind that the restricted specification was regarded as the appropriate way to specify China's grain supply function (see the previous section), discussions of the estimated results below therefore focus on these results (equations 4.1–4.3) although the validity of the restriction imposed is also discussed.

As shown by equations 4.1 to 4.3 in Table 4.4, all the coefficients of the explanatory variables identified have the signs as hypothesized. The positive signs of the coefficients of grain purchasing price, grain growing area and time trend show that grain output responds positively to changes in its purchasing price, growing area and the technological improvement in grain production. Meanwhile, grain output responds negatively to changes in the price ratios of grain/cotton, grain/fertilizer and the climatic variable. These relationships confirm the theoretical predictions. Moreover, the conventional test of significance (t-test) shows all the major regression coefficients (except the intercepts which are discussed separately)

Table 4.4 Total grain output response

Equation	4.1	4.2	4.3	4.4	4.5	4.6	4.7
Intercept	–1,111.70*	2,798.79***	–972.98	–1,684.66*	–1,034.08	–1,032.69	–995.93
	(–1.58)	(7.97)	(–1.06)	(–1.33)	(–1.11)	(–1.12)	(–1.07)
Purchasing price	5.73***	4.67***	3.53***	3.99***	3.73***	2.90***	3.07***
	(6.75)	(3.86)	(4.05)	(3.84)	(4.10)	(2.71)	(2.71)
P.grain P.cotton	–1,494.50***	–1,671.80***	–2,151.69***	–2,101.99***	2,038.25***	1,937.74***	–1,973.79***
	(–6.12)	(–4.59)	(–8.37)	(–7.92)	(–6.96)	(–5.86)	(–5.23)
P.grain P.fertilizer	–1,010.95***	–1,170.44***					
	(–4.52)	(–3.50)					
Area	1.24***		1.35***	1.57***	1.33***	1.35***	1.34***
	(5.92)		(4.96)	(4.15)	(4.84)	(4.95)	(4.84)
Weather	–0.18***	–0.16***	–0.17***	–0.20***	–0.16***	–0.20***	–0.18***
	(6.71)	(–4.08)	(–4.76)	(–3.75)	(–4.53)	–(4.20)	(–4.51)
Time	92.89***	96.43***	82.35***	81.70***	79.00***	86.05***	80.73***
	(15.24)	(10.58)	(11.15)	(10.94)	(9.32)	(10.48)	(10.26)
Dummy variables							
D_1 1959/61					–104.54		153.42
					(–0.83)		(–0.42)
D_2 1962/65						–61.81	
						(–0.82)	
D_3 1966/77						–91.00	–65.12
						(1.02)	(0.76)
D_4 1978/84							36.27
							(0.54)
Price elasticity	0.57	0.47	0.35	0.40	0.37	0.29	0.31
R_2	0.98	0.97	0.97	0.97	0.97	0.97	0.97
F	345.1	180.2	238.6	196.7	196.6	199.4	194.7
D-W	2.23	1.46	1.67	1.56	1.55	1.62	1.55

Notes: Figures in the parentheses are the t-ratios of the estimated coefficients
***Significant at 1% level
*Significant at 10% level

to be significant at the 1% level. The R^2 of between 0.97 and 0.98 indicate that, in spite of its limitations, the model "explains" most of the variation in grain output. This is further supported by the F-test, which is significant at 0.1% level for all the equations estimated and confirms that the model gives a good fit to the data. Generally, the statistical findings are consistent with the description of the behavioural forces at work. It must be emphasized that the specific values of coefficients should not be interpreted directly as supply elasticities as with similar models estimated for Western conditions. The variables employed are used as indicators for general policy decisions. In reality the influences on decisions are much more complex than the simple broad variables might suggest. Details of the major findings are explored below.

Equations 4.1-4.3 indicate that when the grain growing area is included (equations 4.1 and 4.3), it has a significant positive coefficient and makes the intercept negative. However, when the growing area is dropped (equation 4.2), the model produces the anticipated sign and magnitude of the intercept. These results support the idea that the growing area performs the rôle of indicating mainly producer's subsistence requirements (see section 4.1 above).

The signs and magnitudes of the estimated coefficients of grain purchasing price may be interpreted as showing that, except for considerations of self-survival, profitability is indeed a factor of concern to producers in the process of grain supply response. The derived price "elasticities" of output response implies that 1% increase in purchasing price results in a 0.35–0.57% increase in production outcome for the whole period on average. (Note that for the reasons discussed in section 4.1, the calculation of output response in this study takes the purchasing price of the endogenous variable into account only). This means that given the others remain unchanged, one US dollar per tonne increase in real purchasing price can induce about 2.7 to 4.5 million tonnes more grain output. The policy indication of this result is that an increase in purchasing price which makes grain production more profitable can be used as a policy measure to encourage the further development of grain production. However, this result has to be treated with caution for the reasons discussed above.

As anticipated, the two price ratios, grain/cotton and grain/fertilizer, worked well as indicators for the shifts in policy patterns. For equations 4.1 and 4.2 these two variables jointly picked up the effects of the changes in policy patterns on grain production. When the price ratio of grain/fertilizer is omitted as in equation 4.3, its effect is mostly absorbed by the price ratio of grain/cotton. The results are thus consistent with the hypothesis of two alternative policy patterns for the historical development of China's grain sector and that the shifts in policy patterns were a principal cause of the erratic course of the nation's grain production.

The estimated coefficient for grain growing area demonstrates that, holding economic circumstances constant, one more hectare of growing area increases output by 1.3 to 1.4 tonnes of grain. The coefficient on the weather variable indicates that for each additional hectare of the grain crop which suffers from natural disaster there is a loss in production of 1.6 to 1.8 tonnes of grain. The value of the coefficient on the technological improvement variable shows that about 8 million tonnes more grain products are obtained annually from continuing efforts in new technology. These results coincide with expectations, and also provide useful information for future decision-making.

Equations 4.4 to 4.7 are the unrestricted equations which correspond to the restricted equation 4.3 by adding intercept dummy variables for different development periods identified. The validity of the restriction was tested using the F-ratio following Maddala, 1979:

$$F = [(RSS^*-RSS)/(K-K^*)] [RSS/(n-K)]$$

where RSS* is the residual sum of squares of the restricted equation; RSS is the residual sum of squares of the unrestricted equation; K* is the number of parameters in the restricted equation; K is the number of parameters in the unrestricted equation. In this case, the validity of the restriction was tested and accepted at the 5% level of significance. That is, the unrestricted equations performed no better than the restricted version. This result was reinforced by the finding that the estimated coefficients of the intercept dummies in equations 4.4 to 4.7 failed to pass the t test at 10% signficance level. It may thus be concluded that apart from the effects of the two price ratios which were included to depict the shifts in policy patterns, there was no extra significant difference in policy effect on grain supply among different development periods. This could be regarded as further evidence to support the hypothesis that there existed two alternative policy patterns in the historical development of Chinese grain sector.

Despite the general desirability to deflate price data in time-series analysis, the current, nominal, purchasing price was favoured in this analysis. However, for comparison, modelling with deflated prices was also tried. In Table 4.5, equations 4.8 and 4.9 are the equations with deflated purchasing price corresponding to equations 4.1 and 4.3 in Table 4.4 respectively. Generally speaking, the coefficients shown by equations 4.8 and 4.9 have similar magnitude, sign and statistical properties to those in equations 4.1 and 4.3. The exceptions are the estimated coefficients for the grain purchasing price which have increased drastically using the deflated data. There are two practical issues inherent in the model estimation with deflated price. First, due to the large increase in the coefficients for grain purchasing price, the price elasticity derived is 111.5 for equation 4.8 and 38.8 for equation 4.9. This is an unacceptable result notwithstanding the general warning which was given in interpreting the price elasticities derived from this methodology (see section 4.1). Second, the only price index available for deflation is the official series of the gross Retail Price Index. The compilation of this price index, according to the SSB (see *TJNJ 1983*, p592), was a weighted arithmetic mean based on data collection for retail goods prices from certain cities and county towns. The number of commodity prices collected and the number of cities and county towns collecting this data were changing from time to time. In the early 1980s, about 450 and 400 commodity prices were collected by 140 cities and 230 county towns respectively. Therefore not only are there questions about the general accuracy and consistency of the price index compiled, but whether this index is suitable for deflating the grain purchasing price in an analysis of grain producer behaviour is also in serious doubt.

The poor statistics for deflation helps explain the unacceptable price elasticities derived. This unsatisfactory result with deflated prices was repeated in the analyses of individual crops and for this reason these results will not be referred to below.

Given the "two-patterns" hypothesis under test in this chapter, it might be considered that the simple use of zero-one dummy variables could adequately reflect the shifts in the under-lying relationships as the grain economy switched between Pattern A and Pattern B. This should be achieved, for example, by specifying a dummy variable (D = 1) for 1959–61 and 1966–77 when Policy Pattern B was pursued and zero otherwise when Policy Pattern A was implemented. This was tried in two formulations one corresponding to equation 4.3 using the undeflated grain purchasing price and the other corresponding to equation 4.9 with the deflated purchasing price. The results are shown by equations 4.10 and 4.11 in Table 4.5. The general features indicated by these equations are similar to the corresponding equation of 4.3 and 4.9 respectively. The coefficient on the dummy variable is negative and significant, reinforcing the clear message that there is a significant shift in the function between the two periods specified. The Great Leap Forward and the Cultural Revolution are thus associated with a fall in grain output of 2.3 to 2.6 million tonnes per year. However, the critical problem inherent in these results is the poor statistical evidence displayed. The Durbin-Watson test indicates that there is strong evidence for a positive auto-correlation between adjacent error terms (both for equations 4.10 and 4.11). Consequently, the regression coefficients estimated with OLS methods are no longer with minimum variances and "t-" and "F-" tests of significance are invalid.

Such a statistical result was anticipated given prior knowledge about the policy issues. Broadly speaking, it is realized that there are more factors influencing the historical grain production than the single econometric supply equation can cope with. Specifically, it was pointed out in Chapter Two (section 2.6) that the two policy patterns proposed are based on the fact that policy pursued among individual periods share the same basic features, they go in the same broad directions and with similar consequences. Nevertheless, for those periods under the similar policy pattern, for example, the eras of the Great Leap Forward and the Cultural Revolution, there are also differences in how far the policy was taken and the seriousness of the damage created. Therefore, while it is considered as useful and valid to hypothesize the existence of two alternative policy patterns, they are not adequately represented as a strict dichotomy as "policy on" versus "policy off". For these reasons, the characterization of the two policy patterns using the pair of price ratios (price of grain to cotton and the price of grain to fertilizer) was preferred. The evidence from the analysis of each individual grain crop and the two regional analyses was essentially the same as that for the grain sector as a whole.

In estimating the output response equations the statistical technique employed was Ordinary Least Squares (OLS). It is important to be aware of the basic assumptions which are automatically built-in using this technique and the dependence of the results upon the validity of these assumptions. Using multiple regression analysis with time-series data in economic modelling often throws up the twin problems of serial correlation and multi-collinearity. These are discussed next.

Table 4.5 Total grain output response using deflated price series

Equation	4.8	4.9		4.10		4.11
Intercept	−1,019.61* (−1.36)	−350.64 (−0.34)	Intercept	−2,610.52* (−1.54)	Intercept	−2,156.97* (−1.39)
Deflated purchasing price	1,322.46*** (6.24)	460.43*** (2.50)	Purchasing price	2.70* (1.35)	Deflated	397.87 (1.24)
P.grain P.cotton	−1,029.01*** (−3.29)	−2,225.82*** (−7.51)	Area	1.31** (2.46)	Area	1.09** (2.31)
P.grain P.fertilizer	−1,615.95*** (−5.19)		Weather	−0.18*** (−2.90)	Weather	−0.15*** (−2.35)
Area	0.78*** (3.55)	1.09** (3.70)	Time	61.42*** (4.88)	Time	63.63*** (4.29)
Weather	−0.09*** (−3.10)	−0.12*** (−3.11)	Dummy for Policy Pattern B	−232.59*** (−2.25)	Dummy for Policy Pattern B	−265.96*** (−2.93)
Time	84.29*** (11.64)	84.80*** (8.37)				
Price elasticity	111.5	38.8		0.27		33.5
R^2	0.98	0.97		0.92		0.92
F	30.1	181.5		75.2		74.37
D-W	2.21	1.50		0.61		0.60

Notes: Figures in the parentheses are the t-ratios of the estimated coefficients
***Significant at 1% level
**Significant at 5% level
*Significant at 10% level

One of the crucial assumptions of the classical linear regression model is that the disturbances are independently distributed. If this assumption is violated, the disturbances may be auto-correlated or serially correlated (Maddala 1979, p275). In these circumstances the OLS estimates of the regression coefficients are still unbiased but no longer have minimum variance. It also implies that the t- and F-tests of significance for the estimated coefficients and associated confidence intervals are invalid.

The test statistic for the problem of serial correlation is the Durbin-Watson statistic shown as "D-W" in Tables 4.4 and 4.5. Roughly speaking, a D-W value of around 2 indicates the absence of first order auto-correlation of adjacent error terms; values around and below 1 demonstrate the evidence of positive correlation and values around and above 3 suggest the presence of negative correlation in the residuals. The test is usually inconclusive when D-W value lies in the range 1-3. Since the D-W statistics in Table 4.4 are either greater than 1 and less than 2 or between 2 and 3, the Durbin-Watson test is inconclusive at 5% significance level for all the equations displayed. This implies that there is no strong statistical evidence to confirm or deny that there is serial correlation in the unexplained residuals. Given the broad scope of the analysis and its content it was concluded that OLS is a sound enough technique to provide the required results.

Multi-collinearity is a term used to denote the presence of linear or near linear relationships among independent variables. It is a phenomenon inherent in most economic relationships due to the nature of economic magnitudes (Koutsoyiannis 1984, p233). When the explanatory variables are highly correlated among themselves, it is very difficult to assess the unique contribution of each of them. Any statement about an independent variable is contingent upon the other variables in the equation. Consequently structural questions cannot be answered given strong evidence of multi-collinearity.

In the analysis reported above the grain purchasing price and price ratios of grain/cotton and grain/fertilizer were taken as three independent variables in influencing the historical grain output response in China. Although a clear distinction was made in model specification of the separate effects of these variables (section 4.1), this does not rule out the possibility of a strong linear relationship among them. Furthermore, since the estimated results are intended to be used both for clarifying the policy issues as well as the whole pattern of structural relationships, the existence of multi-collinearity was a vital issue for the validity of using OLS and the credibility of policy conclusions reached.

Multi-collinearity can be detected in several ways. First, its presence can be signalled by large coefficients in the correlation matrix. The simple correlation matrices for variables in Table 4.4 and for those in other estimated equations are presented in Appendix Two). However, small correlation coefficients do not always mean there is no multi-collinearity problem (Norusis 1985, p55). Meanwhile since variances of the estimates also increase when independent variables are interrelated, this may result in a significant R^2 for a regression equation and yet few of the coefficients are significantly different from zero (i.e. they have low t-ratios). Furthermore, with highly inter-related explanatory variables, estimated coefficients and associated statistics may be unstable from one model specification to another. Fortunately, all the estimated coefficients of explanatory variables from equations 4.1 to 4.3 are highly significant statistically (high t-ratios), significant R^2 are displayed throughout and the signs and magnitudes of the estimates are stable enough from one equation to another. The combination of these features suggests that there is no strong evidence of the existence of serious multi-collinearity, therefore, results from the estimation are suitable for both prediction purpose and structural analysis. Nevertheless caution should be taken in view of the high correlation coefficients between grain purchasing price, the time trend for technological improvement and the price ratio between grain and fertilizer which are displayed in the simple Correlation Matrix 2.1 of Appendix Two. Bearing all the above discussion in mind it is concluded that equation 4.3 is the best for further empirical application.

4.2.2.2 Rice

The main characteristics of rice production are first, it is the most important grain crop in China for it made up 30% of the grain growing area and 44% of the total output in 1984. Second, due to geological and climatic differentiation in natural conditions, rice is mainly cultivated in south China. In 1984, 13 provinces in the south produced 165.8 million tonnes of unmilled rice product which accounted for 93% of the total output of the nation (*ZNTN 1985*, p49). There is also a strong regional pattern to consumption behaviour; in the south, rice is the favourite and has been the main food grain for the people there for many centuries. Third, among five major grain crops in China, rice has the highest yield (Table 1.8). From 1952 to 1984, the yield of rice increased from 2.41 to 5.37 tonnes/ha, with an annual growth rate of 2.5%. In fact, it is the main beneficiary from the investment in water control, both irrigation and drainage; in improved seed technology; and in the improved availability of chemical fertilizer in south China. Meanwhile, rice growing area has also been increased in the last three decades (Table 1.6). Fourth, rice has been the main grain product to be exported and it is becoming more and more predominant in Chinese grain trade.

In summary, given its importance, the analysis of the rice crop is vital in evaluating the effects of policy. It is expected that the main policy results for the grain sector as a whole will be mirrored for rice production.

The model specification for rice is similar to that used for total grains. There are no separate statistics on the rice area which suffered from natural disaster, thus the aggregate figure used for the estimation of grain sector was employed here as well as for other individual crops later on.

The estimated results are presented in Table 4.6, in which equations 4.15–4.17 are the unrestricted equations corresponding to the restricted equation of 4.13. The validity of the restriction was tested with the F-ratio mentioned and accepted at 5% level of significance. This is supported by the t-tests for the intercept dummies estimated. This evidence confirms that except for the effect from the shifts in policy patterns, there is no extra significant difference among individual development periods in terms of China's rice production during the last three decades. Therefore, the discussions below are concentrated on the results of the restricted equations (equations 4.12 to 4.14).

Generally speaking, the estimated results show that the basic features of the policy impact on rice production is the same as the grain sector. Specifically, several things are worth pointing out. First, in view of the instability of the estimated coefficients of the rice purchasing price and the derived price elasticity, choice has to be made between equations 4.12 and 4.13. Considering the strict requirements of rice crop production conditions, especially the water control system, it is fair to suggest that there is less room for price signals to play a rôle for this crop than the whole grain sector. Equation 4.13 is thus considered to be the appropriate one. Second, the coefficient of the rice growing area variable (0.27) shows a marginal productivity (output per hectare) of just below 3 tonnes/ha, which bears out the high-yielding property of the crop. Third, technological improvement accounts for an

Table 4.6 Rice output response

Equation	4.12	4.13	4.14	4.15	4.16	4.17
Intercept	–213.66 (–0.82)	342.09 (1.23)	–723.90*** (–4.45)	429.25* (1.45)	298.39 (0.86)	428.92 (1.16)
Rice purchasing price	4.98*** (4.26)	1.89*** (1.68)	6.54*** (16.53)	1.70* (1.48)	1.91* (1.66)	0.89 (0.64)
P.rice P.cotton	–564.72*** (–4.49)	–969.3*** (–9.90)	–1,082.52*** (–9.00)	–946.55*** (–9.32)	–935.41*** (–5.03)	–812.26*** (–3.64)
P.rice P.fertilizer	–715.80*** (–4.09)					
Rice area	0.33*** (6.12)	0.27*** (4.16)	0.49*** (9.37)	0.24*** (3.38)	0.28*** (3.75)	0.25*** (3.29)
Weather	–0.08*** (–5.58)	–0.08*** (–4.65)	–0.05*** (–2.46)	–0.08*** (–4.61)	–0.08*** (–3.88)	–0.08*** (–3.40)
Time	23.68*** (5.28)	24.30*** (4.31)		25.36*** (4.38)	24.10*** (4.14)	23.63*** (3.94)
Dummy variables						
D_1 1959/61				–30.00 (0.89)		
D_2 1962/65						–2.07 (–0.04)
D_3 1966/77					–11.45 (–0.22)	44.53 (0.41)
D_4 1978/84						127.47 (1.04)
Price elasticity	0.85	0.32	1.11	0.29	0.32	0.15
R^2	0.98	0.97	0.96	0.97	0.97	0.97
F	321.8	241.8	183.0	200.1	194.4	148.6
D-W	1.74	1.44	1.26	1.34	1.43	1.33

Notes: Figures in the parentheses are the t-ratios of the estimated coefficients
***Significant at 1% level
**Significant at 5% level
*Significant at 10% level

additional 2.4 million tonnes of rice products per year, which makes up 30% of the estimated gain for the whole grain sector. In addition, given the rather high simple correlation coefficients between the rice purchasing price, the time trend for technological improvement and the price ratio of rice and fertilizer (see Matrix 2.2, Appendix Two), the instability of the purchasing price coefficients mentioned above could be caused by the existence of multi-collinearity. Therefore the results estimated must be interpreted and used with caution. Equation 4.13 is considered as the best proxy in serving the purposes of prediction and policy simulation.

4.2.2.3 Wheat

Wheat is the other major grain crop in China. In 1984, 30 million hectares of wheat was grown and 87.82 million tonnes of output produced which accounted for 26.2% and 21.6% of the grain growing area and total output respectively (Tables 1.6 and 1.7). Although wheat can be cultivated in every province of the country, about two thirds of the output comes from the north, and only one third from the south. According to Stone (1989) in 1986 about 18–19% of the wheat area was sown with strong winter habit varieties, about 26–30% were sown with facultative or weak winter habit varieties and the remaining 46–51% were sown to spring habit varieties. The latter were not, as is sometimes thought, confined to the north, but were widely used also in the Yangtze Valley, the Central Plain of Sichuan and the area of east China between the Yangtze Valley and the Huai River. Production is further concentrated on six major provinces: Hebei, Shandong and Henan in the north, and Jiangsu, Anhui and Sichuan in the south, which produced 59 million tonnes of wheat and accounted for more than 67% of the total output in 1984 (*ZNTN 1985*, p50).

Wheat is a principal food grain in the daily life of people in northern China. Because of the importance to national nutritional levels, wheat has benefited from considerable investment in technological improvement. The 4.5% annual yield growth during 1952–84 is the highest achieved among the five major grain crops (Table 1.8). Moreover, the wheat growing area increased 19.4% from 1952 to 1984, despite a net reduction of 8.5% in the total grain area in the same period. Finally, wheat has been one of the few food products to be imported from the world market due to the inadequate domestic food grain supply.

The model specification for wheat is the same as rice as well as the whole grain sector. The results are presented in Table 4.7. As anticipated, the basic features of the supply response mechanism of the crop is the same as for the whole grain sector. First, the validity of the restriction of equation 4.18 (corresponding to the unrestricted equations 4.21–4.23) was tested and accepted at the 5% level of significance. This is further evidence confirming the hypothesis built into the specification. Second, the general characteristics of the results are very similar to those in Table 4.6, that is, the instability of the estimated coefficients for purchasing price, the high marginal productivity of the arable land (2.2–2.86 tonnes/ha), and the high level of technological improvement (1.2–2.4 million tonnes per annum). Among three equations of 4.18 to 4.20, equation 4.20 is considered to be the best, since an elasticity exceeding 1.0 seems too high for a crop which mainly serves self-

Table 4.7 Wheat output response

Equation	4.18	4.19	4.20	4.21	4.22	4.23
Intercept	–3,212.23**	3,237.52***	–2,492.19	–3249.60	–1,844.02	–1,003.59
	(–1.75)	(2.773)	(–1.01)	(1.22)	(–0.49)	(–0.34)
Wheat purchasing price	20.55***	18.91***	8.53*	9.92**	6.29	5.17
	(4.42)	(3.25)	(1.61)	(1.80)	(0.92)	(0.73)
P.wheat P.cotton	–2,060.68***	–2,395.73***	–4,518.49***	–4,545.22***	–3,526.27***	–4,059.11
	(–2.89)	(–2.69)	(–6.74)	(–6.76)	(–2.62)	(–3.66)
P.wheat P.fertilizer	–4,382.38***	–5,331.47***				
	(–4.81)	(–4.82)				
Wheat area	2.20***		2.86***	3.07***	2.37***	2.34***
	(4.07)		(4.07)	(4.16)	(2.02)	(2.66)
Weather	–0.36***	–0.29***	–0.36***	–0.46***	–0.48**	–0.35***
	(–3.50)	(–2.27)	(–2.61)	(–2.62)	(–1.98)	(–2.62)
Time	176.35***	236.86****	119.20****	113.82***	139.25***	128.09***
	(6.64)	(8.58)	(3.73)	(3.50)	(3.26)	(3.80)
Dummy variables						
D_1 1959/61			345.27	120.80		
			(0.94)	(0.24)		
D_2 1962/65				–388.72	–234.47	
				(–0.81)	(–0.80)	
D_3 1966/77				–467.22		
				(–0.91)		
D_4 1978/84					370.55	
					(0.63)	
Price elasticity	1.41	1.30	0.59	0.68	0.43	0.36
R^2	0.97	0.95	0.95	0.95	0.94	0.94
F	176.5	132.2	113.9	94.7	68.2	78.7
D–W	2.09	1.67	1.53	1.53	1.36	1.39

Notes: Figures in the parentheses are the t–ratios of the estimated coefficients
***Significant at 1% level
**Significant at 5% level
*Significant at 10% level

subsistence. However, the elasticity of 0.59 in equation 4.20 which is higher than rice can be justified, as a range of alternative grain and other cash crops compete for the same land.

4.2.2.4 Soybean

Although soybean has always been catalogued in grain production statistically since the founding of the PRC, it is actually an industrial crop for most of the products are industrially processed and consumed as edible oils. Geographically, it is grown mainly in north China, especially in the three provinces of the northeast region (Liaoning, Jilin and Heilongjiang Provinces) which produced 45% of the total output in 1984 (*ZNTN 1985*, p53).

Among the five major grain crops, soybean has the lowest yield. Mainly for this reason, soybean production was not favoured either by the government or producers. The government struggled for an adequate amount of purchased grain to support the industrialization programme and the farmers were more concerned with self-subsistence. The governmental discrimination took the form of 1:1 ratio in soybean purchasing in which, until 1978, a one tonne submission of soybean was accepted as one tonne fulfillment in grain purchasing target. There was also a comparative lack of research and development effort for this crop. The undesirable socio-economic climate was reflected in producers behaviour by all means available to them. This was especially so during the first half of the 1960s, when a serious famine developed. Soybean production was reduced as peasants planted other grain crops to provide their subsistence. These are the major reasons why the soybean growing area was reduced from 11.7 million hectares in 1952 to the lowest figure for some time of 6.7 million hectares in 1976 and has remained below 8 million hectares ever since, leaving production stagnant and below 10 million tonnes (Table 1.6 and 1.7).

In modelling the supply of soybean, two important modifications are made to the general specification. First, since there has been little effort in research and extension activities in this crop, and the general socio-economic climate was unfavourable for the adoption of new technology, no obvious trend for technological improvement is expected. Second, as both corn and wheat are the two major competitive crops for soybean production, two price ratios (soybean/corn, soybean/wheat) were included to reflect this competition. Results from the estimation are presented in Table 4.8.

The results show the expected positive relationship between production and price of soybean. The relationship with cotton is strongly negative. Unfortunately the coefficients on the price ratio variables between soybean and corn and between soybean and wheat are not significant, although they have the expected positive sign. With the exception of the dummy variable for the period 1962/65, the coefficients of the dummy variables were insignificant. It is quite conceivable that when producers were under serious threat of starvation, the first priority was to keep the stomach quiet. Logically, in such times the low-yield crops and crops for non-food products are squeezed. Bearing all these considerations in mind, equation 4.27 is assumed to be the best for further empirical application.

Table 4.8 Soybean output response

Equation	4.24	4.25	4.26	4.27	4.28	4.29
Intercept	501.47** (2.45)	481.09** (2.24)	493.44** (2.28)	481.15*** (2.82)	490.43*** (2.75)	448.90** (2.23)
Soybean purchasing price	1.01*** (4.66)	0.89 (1.13)	0.94** (1.74)	0.91*** (4.99)	0.80*** (2.73)	0.75** (2.37)
P.soybean P.cotton	−217.63*** (−2.83)	−209.95*** (−2.70)	−218.81*** (−2.80)	−189.9*** (−2.91)	−161.21** (−2.23)	−169.32 (−1.26)
P.soybean P.corn		80.77 (0.83)				
P.soybean P.wheat			31.77 (0.34)			
Soybean area	0.48*** (3.74)	0.43*** (2.88)	0.47*** (3.63)	0.49*** (4.87)	0.47*** (4.10)	0.52*** (3.37)
Weather	−0.10*** (−4.75)	−0.10*** (−3.95)	−0.10*** (−4.36)	−0.09*** (−4.99)	−0.08*** (−2.92)	−0.08*** (−2.91)
Dummy variables						
D_1 1959/61					−44.04 (−0.92)	−24.24 (−0.38)
D_2 1962/65				−112.90*** (−3.67)	−128.65*** (−3.26)	−106.46** (−1.84)
D_3 1966/77					−14.00 (−0.28)	20.47 (−0.23)
D_4 1978/84						55.19 (0.48)
Price elasticity	0.39	0.34	0.36	0.35	0.31	0.29
R^2	0.69	0.67	0.68	0.79	0.78	0.77
F	19.0	12.0	14.6	24.6	17.0	14.4
D-W	1.78	1.80	1.78	2.20	2.30	2.22

Notes: Figures in the parentheses are the t-ratios of the estimated coefficients
***Significant at 1% level
**Significant at 5% level
*Significant at 10% level

4.2.2.5 Other grain crops

Due to the limitations imposed by the available data, grain crops other than rice, wheat and soybean are aggregated in this analysis. As shown by Table 4.9, corn and potatoes (sweet potato and white potato) are the two main crops within this category which made up 36% of the total growing area of other grain crops in 1952 and 64% in 1984. In terms of output, these two crops accounted for about half of the total output of other grain crops in 1952 and this increased to more than three quarters in 1984. Corn and potatoes are high-yielding crops with 3.95 and 3.17 tonnes per hectare in 1984 respectively, while the yield for others is much lower, less than 2 tonnes on average in 1984. Nevertheless, all of them show a considerable yield improvement during the last three decades.

Geographically, most of the crops are grown in North China. For example, 77.3% of the corn, 92.1% of sorghum, 99.4% of millet were produced in the north in 1984 (*ZNTN 1985*, pp52–53). Production of these three crops is further concentrated in six provinces: Shandong, Hebei, Henan, Liaoning, Jilin and Heilongjiang. For tuber crops, there exists a rather equal distribution between south and north. However, four provinces can also be identified as the major growing regions: Shandong, Sichuan, Henan and Anhui, for they produced 54% of the total output in 1984 (*ZNTN 1985*, p51).

From the consumption point of view, products of most crops in this category are less preferred than food grain for direct consumption. Historically, when per capita grain availability was low, a greater proportion of the products were substituted for food grain; but when per capita grain availability was higher, more of these products were allocated for other purposes, especially as feed grain for livestock industry.

The policy variables for these crops appear to have a less clear cut effect than those estimated for other crops. This may partly be due to the lack of homogeneity

Table 4.9 The production situation of other grain crops

Items	Growing area (10,000 ha)		Yield (tonne/ha)		Output (10,000 tonnes)	
	1952	1984	1952	1984	1952	1984
Total	5,914	4,284	1.15	3.07	6,784	13,153
In which:						
Corn	1,257	1,858	1.34	3.95	1,685	7,341
Potatoes	869	899	1.88	3.17	1,633	2,848
Others	3,788	1,527	0.91	1.94	3,466	2,964

Note: Potatoes refers to the combination of sweet potato and white potato. Other grain crops include sorghum, millet, barley, oats, rye and legumes other than soybean.

Source: Derived from Table 1.6, 1.7 and 1.8.

of the group itself. Although similar responses to the changes in policy patterns can be expected, there has been a general tendency to discriminate against these products for most of the period studied.

Results from the estimation are shown in Table 4.10. The most interesting evidence is that in the F-tests for the validity of the restrictions between the equations 4.30 to 4.32 and the corresponding unrestricted equations 4.33 to 4.35 the restrictions were rejected at the 5% level of significance. This is confirmed by the high significance of the coefficient of the dummy variables all of which passed the t-test at the 5% significance level. All this suggests that there were significant differences in policy impacts on the production of other grain products between the different development periods of the last three decades. The reasons for this could be, first, as it is a mixture of different crops, problems could be created by aggregation error. Second, the high-yielding property of corn and tubers and their adaptability to a wide range of natural environments favoured crops in this category during the early stages. However, the continuing improvement in production conditions and intensive research effort made on the more mainstream crops, such as rice, wheat and corn, plus consumer's preference for fine grain, made it possible and reasonable to shift from the crops in this category (especially tubers) to those more favoured crops. Nevertheless, despite these considerations, it appears that this category of cereals also provides general support for the hypothesized policy patterns.

4.2.3 Regional output response estimation

4.2.3.1 Pindu County of Shandong Province

Shandong is the biggest agricultural province in north China. In 1984 it produced 30.4 million tonnes of grain (of which 12.8 million tonnes were wheat, 9.9 million tonnes corn and 5.3 million tonnes tubers), 1.725 million tonnes of ginned cotton and 1.82 million tonnes of oil-bearing products (of which, 1.8 million tonnes were peanut) which accounted for 7.5%, 27.6% and 15.2% of the outputs of the whole nation respectively (*ZNTN 1985*, pp48–52).

Pindu is a county located in the northeast part of Shandong Province. It is the largest county for agricultural production and has a similar farming composition as the province as a whole. The basic features of the county's farming system are shown in Table 4.11.

The main developments in grain production in this region demonstrate a similar picture as that for the whole nation (see Chapter One). First, from 1952 to 1984, total grain output increased considerably despite a decrease in growing area (see Appendix Table 1.4). Second, while a significant improvement in yield has been achieved for all the grain crops, the growth in output is partly due to the shift in growing area from low-yielding to high-yielding crops; in this county, mainly from millet, sorghum and soybean to corn (Table 4.11). Third, a zig-zag course is displayed both by the grain and agricultural production generally during the last three decades. All the features presented suggest that this county provides a typical example in terms of policy impact on the grain production in the region.

Table 4.10 Other grain output response

Equation	4.30	4.31	4.32	4.33	4.34	4.35
Intercept	5.33 (0.15)	0.65 (0.12)	15.18 (0.34)	–49.14* (–1.50)	–66.77** (–2.08)	–23.16 (–0.50)
Purchasing price	0.41*** (4.09)	0.11*** (2.54)	0.08** (1.74)	0.25** (2.08)	0.07 (1.27)	0.11* (1.44)
P. other grain crops P. cotton	–4.17 (–0.28)	–39.80*** (–3.42)	–30.80*** (–2.51)	–18.28* (–1.08)	–24.91* (–1.45)	–44.78** (–1.80)
P. other grain crops P.fertilizer	–94.43*** (–3.20)			–48.56* (–1.67)		
Area of other grain crops	0.01 (1.17)	0.01** (1.95)	0.01 (1.30)	0.02*** (3.20)	0.02*** (4.45)	0.02** (2.20)
Weather	–0.01*** (–3.94)	–0.01*** (–2.62)		–0.01*** (–5.27)	–0.01*** (–5.49)	
Time	3.09*** (7.53)	2.32*** (6.02)	2.14*** (5.12)	4.26*** (8.87)	4.22*** (8.48)	2.83*** (4.44)
Dummy variables						
D_1 1959/61				–0.01*** (–2.60)	–0.09*** (–2.61)	–0.05 (–1.00)
D_2 1962/65				–17.14*** (–3.75)	–21.10**** (–5.21)	–11.95** (–2.18)
D_3 1966/77				–21.61*** (–2.93)	–28.14*** (–4.33)	–4.98* (–1.67)
D_4 1978/84				–19.55*** (–2.09)	–23.35*** (–2.49)	–8.25 (–1.12)
Price elasticity	1.10	0.30	0.22	0.68	0.19	0.31
R^2	0.92	0.89	0.87	0.95	0.95	0.89
F	62.6	54.4	54.8	64.8	66.6	32.2
D–W	1.81	1.20	1.35	2.11	1.78	1.82

Notes: Figures in the parentheses are the t–ratios of the estimated coefficients
***Significant at 1% level
** Significant at 5% level
* Significant at 10% level

Table 4.11 The production composition of the major farming crops in Pindu County, for selected years

Year	Wheat		Corn		Tubers		Millet	
	Area (10,000 ha.)	Output (10,000 tonnes)	Area (10,000 ha.)	Output (10,000 tonnes)	Area (10,000 ha.)	Output (10,000 tonnes)	Area (10,000 ha.)	Output (10,000 tonnes)
1952	9.31	6.71	1.30	1.83	4.23	6.75	3.14	4.14
1957	9.71	7.79	3.87	5.05	4.52	7.36	1.68	1.89
1962	7.82	4.11	1.97	2.48	5.93	9.02	1.47	1.25
1965	8.21	6.16	2.70	4.41	5.69	9.64	1.00	0.95
1975	8.08	10.90	4.01	10.31	5.63	22.53	0.75	1.02
1982	7.83	17.04	5.42	25.61	3.25	15.10	0.11	0.29

Year	Soybean		Sorghum		Cotton		Peanut	
	Area (10,000 ha.)	Output (10,000 tonnes)	Area (10,000 ha.)	Output (10,000 tonnes)	Area (10,000 ha.)	Output (10,000 tonnes)	Area (10,000 ha.)	Output (10,000 tonnes)
1952	6.06	4.00	3.45	4.58	1.33	0.34	1.67	2.65
1957	4.91	2.54	1.29	1.16	2.05	0.31	2.14	1.96
1962	3.69	1.72	2.04	1.12	1.53	0.13	1.36	1.28
1965	3.63	2.39	1.03	0.84	1.90	0.60	1.75	2.83
1975	1.61	1.39	0.51	0.66	1.80	1.35	1.87	2.38
1982	0.93	1.77	0.14	0.30	3.13	1.69	2.26	5.53

Notes: Figures for grain output are for unmilled products.
Figures for cotton output are for ginned cotton.

Source: From *Report on the Agricultural Zoning of Pindu County 1984 (Pindu Xian Nongye Quhua 1984*, hereafter*PNQ 1984*) pp200-201.

Supply equations were estimated for the county data in the same manner as for the national data. The results are summarized in Table 4.12. The results demonstrate a similar pattern of policy impacts on grain production as for the whole nation. Again the F-tests on equations 4.36 and 4.37, plus the t-test on the coefficient of the dummy variables in equations 4.38 to 4.41, demonstrate that little additional explanation is yielded by differentiating the period by dummy variables. The remaining discussions of the empirical results therefore pertain to the restricted equations. Specifically, the low marginal productivity of land (0.38 to 0.83 tonnes/ha derived from equations 4.36 and 4.32) partly reflect the characteristics of the resource endowment in this county, and is partly due to the coexistence of a reduction in total growing area on the one hand and a shift from low- to high-yielding crops on the other. The strong trend of technological improvement in this region is shown by the estimated 17,500 to 22,700 tonnes annual increase in grain output. This is quite consistent with the prior knowledge of the tremendous efforts

Table 4.12 Pindu County grain output response

Equation	4.36	4.37	4.38	4.39	4.40	4.41
Intercept	567.88*** (4.30)	715.77*** (5.29)	562.50*** (4.00)	441.60*** (2.78)	744.27** (5.35)	542.53*** (2.98)
Grain purchasing price	1.27*** (2.79)	0.67* (1.48)	1.11** (2.22)	1.38** (2.61)	0.64 (1.19)	0.54 (1.01)
P. grain / P. cotton	–506.26*** (–3.74)	–747.95*** (–6.41)	–495.18*** (–2.68)	–319.36* (–1.70)	–761.01*** (–5.45)	–496.16*** (–2.37)
P. grain / P.fertilizer	–355.06*** (–2.81)		–366.55*** (–2.98)	–402.37 (–3.12)		
Grain area	0.17 (0.70)	0.37* (1.39)	0.17 (0.63)	0.20 (0.74)	0.34 (1.24)	0.27** (0.85)
Time	22.73*** (5.70)	17.55*** (4.45)	22.39*** (5.39)	18.39** (3.56)	17.50*** (4.16)	17.00*** (2.83)
Dummy variables						
D_1 1959/61			–56.30 (–1.28)	–58.99* (–1.36)	–48.34** (–1.34)	–89.69** (–1.32)
D_2 1962/65				–43.49 (–0.73)		–95.50* (–1.43)
D_3 1966/77				21.39 (0.26)	4.27 (0.13)	–74.32 (–0.83)
D_4 1978/84			26.37 (0.48)	65.49 (0.56)		–53.82 (–0.41)
Price elasticity	0.84	0.44	0.73	0.91	0.42	0.35
R^2	0.91	0.88	0.91	0.91	0.88	0.89
F	62.6	61.1	48.9	40.3	41.3	32.3
D-W	2.02	1.69	2.51	2.53	1.89	1.74

Notes: Figures in the parentheses are the t-ratios of the estimated coefficients
***Significant at 1% level
** Significant at 5% level
* Significant at 10% level

made in improving the production conditions (see *PNQ 1984*). It is a critical indicator which explains why a total growth rate of 280% (from 0.284 million tonnes in 1952 to 0.794 million tonnes in 1984) was achieved despite the 36.4% net reduction in growing area (from 0.279 to 0.177 million hectares). Meanwhile, the comparatively high price elasticity of supply (0.44 to 0.84) indicates the sensitiveness of price signal in influencing the farming structure when rather high per capita grain output was obtained and alternative crops (mainly cotton and peanut) compete for the limited amount of arable land. Finally, in view of the high correlation coefficients among three independent variables displayed (see Matrix 2.6, Appendix Two) and the instability of the estimated coefficients between equations 4.36 and 4.37, the existence of multi-collinearity must be suspected.

4.2.3.2 Jianou County of Fujian Province

Jianou County of Fujian Province is a typical grain growing area in south China in which rice is the predominant crop in grain production. In 1981, 56,300 hectares of rice was cultivated in the region which made up 92% of the total grain growing area and 78% of the total crop farming. In output terms, 218,039 tonnes of paddy rice was produced which accounted for 97% of the grain output in total (see the Appendix of *Report on the Comprehensive Agricultural Zoning of Jianou County 1984* (*Jianou Xian Zonghe Nongyue Quhua 1984* (hereafter *ZNQ 1984*)). Although arable land made up only 8.3% of the county's territory, 62.6% of the gross value of the agriculture sector was generated from the farming sector. For such a one-sided crop system which also demonstrates a zig-zag course in its historical development (see Appendix Table 1.4), it is interesting to investigate whether the pattern of policy impacts is the same as for the whole nation. According to local producers and county officers, it is a region that has always fixed the policy maker's eye because of its importance as a major source of exports of grain products to other regions of the province and because of the profound impact policy has had on grain production for a very long time.

Equations 4.44 to 4.47 in Table 4.13 are the unrestricted equations corresponding to the restricted equation of 4.43, the validity of the restriction is justified by the F-test at 5% level of significance, and by the evidence from t-test for the intercept dummies estimated. Therefore the restricted equations in Table 4.13 are the main focus in the result assessment. As shown by equations 4.42 and 4.43, the basic picture of policy effects in this region is identical to Pindu County, the typical grain growing region in the north, and the nation as a whole. Moreover, all the signs and magnitudes of the estimated coefficients are reasonable, the statistical tests are robust and the estimated equations are stable. These are the most satisfactory results of supply estimation in this chapter.

4.3 Summary

A set of important issues concerning the supply response behaviour of China's grain sector since the founding of the PRC have been clarified in this chapter.

Table 4.13 Jianou County grain output response

Equation	4.42	4.43	4.44	4.45	4.46	4.47
Intercept	72.50***	72.35****	77.86**	39.43	51.27	40.87
	(1.80)	(1.83)	(1.90)	(0.91)	(1.22)	(0.93)
Grain purchasing price	0.34***	0.34***	0.33***	0.28***	0.27***	0.29***
	(3.72)	(3.83)	(3.67)	(3.18)	(3.03)	(3.25)
P. grain / P. cotton	−176.83***	−179.09***	−182.95***	−135.28***	−142.38***	−141.69***
	(−6.95)	(−7.49)	(−6.78)	(−3.84)	(−4.68)	(−4.00)
P. grain / P.fertilizer	−4.65					
	(−0.30)					
Grain growing area	1.01***	1.03***	1.03***	1.10***	1.03***	1.10***
	(3.54)	(3.69)	(3.64)	(4.06)	(3.74)	(4.01)
Time	4.22***	4.02***	4.56***	3.01***	3.86***	3.16***
	(3.46)	(4.06)	(3.98)	(2.72)	(3.03)	(2.83)
Dummy variables						
D_1 1959/61			−7.50	−9.29*	−11.27*	
			(−1.04)	(−1.35)	(−1.03)	
D_2 1962/65			1.87	−1.39		0.97
			(0.27)	(−0.21)		(−0.07)
D_3 1966/77					−9.19	
					(−1.07)	
D_4 1978/84				20.35	6.21	18.50
				(0.97)	(0.39)	(0.78)
Price elasticity	0.48	0.48	0.48	0.40	0.38	0.42
R^2	0.96	0.96	0.96	0.97	0.97	0.97
F	169.6	219.1	142.7	136.5	142.6	154.0
D-W	1.59	1.60	1.69	1.78	1.70	1.64

Notes: Figures in the parentheses are the t-ratios of the estimated coefficients
***Significant at 1% level
**Significant at 5% level
*Significant at 10% level

First, evidence shows that with the conditions of low productivity and low per capita grain availability, self-subsistence is a fundamental issue which concerns the Chinese grain producers. From the regional models, the positive signs and appropriate magnitude of the intercept coefficients estimated tell the story. In the national response equations, this effect has either been picked up by the included grain growing area or by the intercept directly.

Second, all of the estimated equations demonstrate a positive relationship between price (usually the purchasing price of the endogenous variable) and the corresponding output. This suggests that, beyond the motivation of self-subsistence, profitability is a factor which influences producer's supply response behaviour.

Third, the most important result from this elementary supply modelling is that the general hypothesis which was described in the chapters above has been empirically confirmed. In particular the results confirm that there were two alternative policy patterns, which can explain the general success or failure among different development periods, that is the zig-zag course of the historical grain production. This is confirmed by the oscillation function performed by the price ratios manipulated by the government (price ratio between grain and cotton and of grain and fertilizer), the restrictions imposed (justified by the F-test) and the statistical insignificance of the estimated intercept dummies (except in the case of soybean and other grain crops).

Fourth, the estimated coefficients on the time trend variables which were specified as indicators of technological improvement in grain production (not in the case of the soybean crop) has reflected reasonably well the continuing efforts made in improving the production conditions. Meanwhile, results also show the changes in natural environment, especially changes in the weather, are still an important obstacle to be overcome if a stable and accelerated growth of the nation's grain production is to be achieved.

Although the theoretical hypothesis of the policy patterns proposed in section 2.6 has been confirmed in this chapter, to present full answers for questions (a) and (b) posed in the Introduction raises further issues. For example, it is important to ask how the policy patterns worked and why it was successful in one period but failed in another. Other factors determining the Chinese peasant farmers' supply response behaviour, cannot be directly derived from sectoral modelling which is at too high a level of aggregation. More solidly based conclusions can only be obtained after further detailed investigation. These remaining issues are revealed in the next chapter.

Chapter Five

Supply Response Analysis – Discussions and Conclusions

Aggregate supply analysis can only provide the broadest indication of the complex web of interactions between farmers and the political, economic and institutional setting in which they operate. This chapter will explore issues of peasant farmers' supply response behaviour under the centralized management system in China. It will focus on the policy mechanism under which agriculture was developing, the internal and external linkages of the agricultural sector and the policy impacts on agricultural production. The chapter concludes with answers to the first two questions raised in the Introduction and a brief summary of the major lessons from the historical development of China's agricultural production.

5.1 Chinese Peasant Farmers' Supply Response Behaviour

5.1.1 Introduction

Supply response behaviour of the peasant farmers in developing countries has been a popular research topic for many development economists. Discussions have mainly been focussed on the issue of producers' responsiveness to the price signals. According to Seini (1984, p78), two categories of prevailing hypotheses can be identified: (i) that peasant farmers respond quickly, normally and efficiently to relative price changes; and (ii) that institutional constraints are so limiting that any price response is insignificant.

The well-known proponent of the first hypothesis is Schultz (1964, p33) who argued that like the commercial farmers in modernized agriculture, the rate at which peasant farmers of the traditional agriculture accept a factor of production depends on its profit with due allowance for risk and uncertainty. Such a proposition has been supported by many researchers with a number of empirical studies. As a proponent of the second hypothesis, Nerlove (1979, pp883–885) argued that increased production in peasant agriculture is more dependent upon factors other than prices, such as infrastructural constraints, technical change, demographic change and government intervention. Specifically, "one might infer little supply response to price observed in central markets, for example, simply because such prices are largely irrelevant to the allocation problems which these farm households resolve."

Although a positive relationship between grain purchasing price and output response was found to exist as described in Chapter Four, given the rather aggregate analysis and the fact of the coincident movements between the purchasing price and changes in policy patterns, more evidence has to be presented before a convincing conclusion about Chinese peasant farmers' behaviour can be derived. This is the subject of the following section.

5.1.2 The peasant farmer as a profit maximizer

5.1.2.1 The peasant farmer before the founding of the PRC

China is a nation with 4,000 years of history in agricultural civilization. The requirement for being a good farmer over 2,000 years ago was given by the *Book of Filial Piety* as "Work according to seasons/Suit crops to soils for profit/Guard behaviour spend wisely/Nurture parents with honor" (quoted from Buck 1930). After analysing relevant evidence concerning peasant farmers' supply behaviour in China before the founding of the PRC, researchers unanimously agreed that the traditional Chinese farmer is an economically rational producer. Tawney (1932, p51) described the Chinese peasant as "a highly skilled farmer, who has achieved ... an extraordinary efficiency". The conclusions by Perkins (1966, p24) is that "the first and most essential feature is that the China peasant farmer, in a fundamental sense, has generally been economically rational. Within the scope of his limited knowledge and given the high risks involved in almost any agricultural undertaking in China ... he has attempted to maximize his expected return". This argument was echoed by Lardy (1983, p11) who claimed that "In most respects the behaviour of Chinese farmers appears to be consistent with the view advanced by Theodore W. Schultz and others that peasants respond efficiently to new production and marketing opportunities". These conclusions are supported mainly by evidence of farmers' behaviour in land utilization, production innovation and agricultural marketing.

Price signals and land utilization In his study of 16,786 farms in 168 localities and 38,256 farm families in 22 provinces in China, 1929–33, Professor Buck (1937, p347) found that a rise in farm prices before 1931 stimulated agricultural production and increased the intensity of land usage. This was demonstrated by reclamation of new lands and use of more fertilizer and labour in crop farming. On the other hand, the fall in prices after 1931 had an opposite effect of curtailing agricultural production. He therefore concluded that price signals "have a very important bearing on the intensity of the utilization of the land".

Peasants' responsiveness in production innovation Unlike Western agricultural production, increased sophistication of Chinese farming was not evident in mechanical technology, changes in the pattern of cropping were the primary source of technical change during the Ming (1368–1644) and Ching (1645–1911) dynasties (Ho 1959, p169). These changes were mainly induced by the introduction of improved seed varieties and new crops from abroad (Lardy 1983, p8). The typical example of the improved seed varieties was early-ripening rice (Ho 1956). These varieties had a shorter growing period and lower water demand, properties which allowed intensification. Thus the cropping pattern for rice production changed, allowing double-cropping of rice in central China and the spread of rice cultivation to higher lands with less adequate water supply. These seeds were initially planted in southeast China but were grown widely in Zhejiang, Jiangsu, Jiangxi and Fujian Provinces within two centuries and were further diffused to the southwest and the central region of China during the Ming dynasty (Ho 1959, p171). The adoption of new crops such as corn, potatoes, cotton, peanuts and tobacco crops, which were mostly introduced during the Ming and Ching dynasties and successfully spread to those appropriate regions, had also led to far-reaching changes in cropping systems (Ho 1955, 1959). These observations confirmed Tawney's argument that "given ocular demonstration of the practical advantages of, for example, better seeds", Chinese peasant farmers are "quick to take advantage of them" as long as their means allow (Tawney 1932, p90).

The development of commercialized production In his study of 2,866 farms in 17 localities and seven provinces in China in the late 1920s, Buck (1930, pp196–199) found that an average of 53% of the value of farm products was sold off-farm, within which the proportion for north China is over 40%, and that of east central China is more than 60%. Buck's conclusion was that there existed a significant commercialized orientation in agricultural production and his interpretation was that "the farmer must have considerable need for ready cash to meet his various requirements" (Buck 1930, p197). Although his sample was biased upward, as pointed out by Perkins (1966, p25), it leaves little doubt of the importance of the market to the Chinese farmer in general. Perkin's own conclusion was, "before the communists came to power, Chinese farmers were heavily orientated toward the market and significantly influenced by prices paid for their produce" (Perkins 1966, p28).

5.1.2.2 The peasant farmer under the centralized management system

Despite Lardy's (1983, p31) general statement that peasants' responsiveness to price signals in the early 1950s was unchanged (1983, p31), the hypothesis that the peasant farmer acts as a rational producer under the centralized management system has not been effectively explored. This will be examined below.

Price signals and grain production and procurement since 1978 From 1978 to 1983, total grain output in China increased from 305 to over 387 million tonnes, with a total growth rate of 27.1%, which drove the per capita grain output of the rural population from less than 195 up to 235 kg, a 21.4% improvement. During the same time period, the amount of grain purchased by the government increased from 50.7 to 102.5 million tonnes, a 102% increase. The much higher growth in procurement partly reflects the increased marginal propensity of the rural households to trade grain after meeting self-subsistence (in the Chinese case, per capita output of 200 kg is generally agreed as the turning point). However, more important is the large increase in the grain purchasing price. As discussed in Chapter Two, the grain purchasing price for compulsory quota was raised 20% in 1979. At the same time an extra 50% price premium was also offered for above-quota submission as an incentive for farmers to produce and sell more grain products. In practice, it is the above quota price which attracted producers to sell more products. Statistical data shows that during 1978–83, the amount of purchased grain which attracted the above-quota premium increased from 12.15 million tonnes to more than 55 million tonnes while the fixed quota purchasing reduced 2.15 million tonnes at the same period. As a result, the percentage of above-quota purchasing in total grain procurement rose from 34.5% in 1978 to 72.4% in 1983, and the government budget spending in grain subsidy correspondingly increased from less than 1 billion to 7.3 billion yuan (Ji 1984, p49). For this reason, the two track pricing and purchasing system was changed to a fixed ratio system. That is, 30% would be bought at the original quota price and the remaining 70% at the premium price (Chen 1985, p15). A similar picture is also displayed by the purchasing of agricultural and sideline products as a whole in recent years, which is presented in Table 5.1.

After several years of continuing growth in grain production, 1985 showed a net reduction of 27 million tonnes in grain output (Ho 1986, p1). Apart from the effect of weather conditions, this set-back was mainly due to the reduction in growing area and peasants' enthusiasm for grain production caused by the increase in industrial inputs price which pushed the cost up and made grain growing less attractive. In the agricultural minister's language, it is "due to the fundamental reason of the rôle played by the law of value" (He 1986, p2); that is, the profitability as seen by producers. This view is shared by a number of empirical studies such as Investigation Group of the Chinese Academy of Agricultural Science (1985), Zhong (1985) and Zhao and Ming (1986).

Farmers' supply response in cash crop production Not only do peasant farmers respond to price signals quickly in grain production, but also, they respond to price changes in cash crop production. Because cash crops are in any case more

Table 5.1 Total purchasing of agricultural and sideline products and its composition (in billion yuan)

	Total purchasing	In which quantities under different purchasing systems were			
		Fixed price premium	Above-quota	Negotiated price	Market price
1978	55.79	47.24	4.42	1.02	3.11
1983	126.50	60.70	35.50	17.00	13.30
% of increase from 1978 to1983	126.7	28.5	703.2	1,566.7	327.7

Notes: The trading under market price are made in free market between farmers and urban residents, others are between farmers and official marketing system. All figures are in current prices.
Source: *ZCWTZ 1952-1983*, p112.

market orientated this response may be more price elastic. Cotton is the most important cash crop of the nation, both because it is the main source for clothing one billion Chinese people and due to the critical rôle played by the textile industry in the foreign trade. However, from 1952 to 1977, the purchasing price for ginned cotton stagnated around 4,000 yuan per tonne for 16 years, and for most of this time the price ratio between cotton and grain was fixed at a level well below 1952 (Appendix Table 1.3). The cotton growing area was thus reduced from 5.58 million hectares in 1952 to 4.84 million hectares in 1977 and output stagnated at around 2 million tonnes for 20 years between 1958–77 (*TJNJ 1983*, pp155,159). Since the early 1970s, an increase in cotton production was urged every year by the top authorities but got little response from the producers. A survey by Chen (1978) in 1978 in Hebei Province, one of the major cotton growing areas in north China, the high pressure on grain procurement and unprofitability of cotton farming, encouraged farmers to take several measures to discriminate against cotton production. First, the real growing area for cotton crops was lower than it was declared to be (which usually is identical to the target allocated to farmers); second, land for cotton growing was interplanted with wheat, corn and other crops. These two factors suggest that the real growing area for cotton was 30% less than that of the planned target. Third, farmers allocated their marginal land to cotton production and also a considerable proportion of fertilizer supplied by the government for cotton was used for grain production. Finally, labour use on cotton was also discriminated against in terms of quantity, quality and timeliness.

This was a common phenomenon at that time for most of the major cotton producing regions of the nation. In Shanxi Province, the usual means of discrimination were by pushing the cotton growing to the high land and using less fertilizer for the crop. Nevertheless, mainly due to the significant increase in its purchasing price, cotton production was booming in the early 1980s with the output increased from 2.17 in 1978 to 6.26 million tonnes in 1984 (Table 1.5). In 1983, state purchasing of cotton reached 4.5 million tonnes. After deduction of 3.25 million tonnes for domestic processing and 50,000 tonnes of net export, 1.2 million tonnes of ginned cotton was piled up in one year (He 1984, p4). In 1984, the overstocking condition was even more serious due to the unprecedented harvest achieved. Much trouble was created because there was insufficient storage capacity and the stocks represented a great financial pressure on the government. As a result, the pricing policy for cotton procurement was also adjusted and the 5% extra subsidy for cotton production in the north was withdrawn (Qiao 1985, p26).

Another example, which merits discussion is tobacco production. While a considerable rise in purchasing price for the major agricultural products was made in 1979, the purchasing price for cured tobacco remained unchanged. Production was therefore curtailed by producers and output dropped from 1.05 million tonnes in 1978 to 0.72 million tonnes in 1980. However, following a substantial increase in purchasing price in 1981, 1.2 and 1.85 million tonnes of output were registered for 1981 and 1982 respectively. This changed the shortage in domestic tobacco supply to a big surplus within two years (*TJNJ 1983*, pp 160,480).

Although only two typical crops are examined here, a similar situation prevails for the production of many other cash crops. A general lesson since 1978 is that as long as grain supply for producer's self-subsistence is available, farmers choose to grow those crops which show the highest potential profitability. Knowing the Chinese farmers supply response behaviour, it is possible to understand the increase in the productions of cotton, oil-bearing and sugar-yielding crops by 189%, 128% and 101% respectively between 1978–84 (Table 1.5).

The development of rural industry Since diversified development of the rural economy has been encouraged, starting in 1978, rural industry run by the basic production units (for the definition, see section 1.3) has been flourishing. From 1980 to 1984, the gross value of rural industry increased from 24.8 to 57.7 billion yuan, with a 23.5% average growth rate per annum (in 1980 yuan) (*TJNJ 1985*, p239). Its proportion in the total gross value of agricultural output rose from 11.2% to 17.1% during the same period (*TJNJ 1985*, p241). This growth in rural industry was partly promoted by the serious labour surplus in the rural area, but more importantly by the attraction of high profits in rural industry given the existence of big price differentials between agricultural and industrial products.

Wuxi in east China's Jiangsu Province is the most famous area for rural industrial development of the nation. According to Professor Xue (1985, pp18–20), output value of the rural industry in the area made up 40% of the area's total output in 1984, and 40% of the rural employees were engaged in industrial production. Although the equipment and technical capacities of the rural factories are generally inferior to urban sites, and raw material consumption, especially energy consumption, is higher than those of the urban state-owned factories, it is still more profitable than farming. In order to balance the higher incomes of the factory employee with those of farmers, the factory-tied households have to pay the specialized grain growing households several hundred yuan for them to work each hectare of land a year. Moreover, some township governments (People's Communes in the early days) also provide grain producers with 200 yuan per tonne for grain sold to the state. Despite these increases in cost, the rural industry in this area will continue to grow at a very high rate in the near future. Under such conditions, the conclusion from Professor Xue is, "Without subsidies, few, if any peasants would specialize in growing grains, while others would only grow enough to feed themselves" (Xue 1985, p19). Nowadays, subsidizing grain production with revenue from the industrial sector within the basic accounting units is regarded as a major policy issue by the government for protecting grain production.

Despite the profit maximizing behaviour of the peasant farmers, which has led to considerable production specialization, significant commercialization and diversified rural development as described above, self-subsistence remains a fundamental dimension in determining Chinese farmers' supply response behaviour. This is discussed next.

5.1.3 The peasant farmer as a self-subsistence producer

As shown by Table 5.2, considerable proportions of major agricultural products (except for cotton) were still used for producers' self-subsistence until 1984. In terms of the total value of agricultural output, less than half was traded. For the grain products, only about one third was marketed. After deducting the quantities sold back to rural areas, the net trading of grain was less than one fourth, i.e. 23.2% in 1984 (*TJNJ 1985*, p482). The main reasons for the self-subsistence nature can be summarized as follows.

Low labour productivity China has a high population to land ratio and a less developed industrial sector. Up to 1984, three fourths of the labour was still predominantly engaged in agricultural production. The labour productivity is particularly low as shown by Table 5.3. In 1978, output per unit of labour of grain and meat (pork, beef and mutton) was 979 and 27.5 kg respectively, which is only 42.7% and 16% of the world average of the same year (*Statistical Collections of Overseas Farming and Animal Husbandry 1978–1979* (hereafter *SCOFA 1978–1979*)). Low labour productivity is the first reason for the less developed commercial production.

Weakness in social infrastructure The weakness in social infrastructure is exemplified by a lack of marketing facilities and services, that is, processing and

Table 5.2 Marketing rates of major agricultural products

	Unit	1978	1980	1982	1984
Total purchasing of agricultural and sideline products	billion yuan	55.89	84.22	108.30	144.00
Proportion of the total value of agricultural output	%	39.9	43.8	43.6	45.3
In which:					
Grain purchasing	million tonnes	61.74	73.00	91.86	141.69
Proportion	%	20.3	22.8	25.9	34.8
Purchasing of ginned cotton	10,000 tonnes	204.30	268.13	349.66	590.00
Proportion	%	94.3	99.1	97.2	94.3
Purchasing of edible oils	10,000 tonnes	115.5	194.5	308.0	322.5
Proportion	%	55.9	71.2	71.9	67.4
Purchasing of pigs	million heads	109.37	142.50	144.63	152.39
Proportion	%	67.9	71.7	72.1	69.1
Purchasing of aquatic products	10,000 tonnes	315.0	280.0	336.8	360.4
Proportion	%	67.7	62.3	65.3	58.2

Notes: All the purchasing figures are measured in production year (April-March). Total purchasing in value term are in current prices. The trading figures take neither the selling and buying activities within the rural area nor the selling back to the countryside by the official marketing system into account.
Source: *NCTNJ 1985*, p127.

Table 5.3 Labour productivity of the agricultural sector

	Unit	1978	1984	% increase during 1978-84
Total gross value of agricultural products	yuan/person	448.8	883.7	96.9
Grain output	kg/person	979.1	1,131.7	15.6
Ginned cotton	kg/person	6.9	17.4	150.4
Oil-bearing products	kg/person	16.8	33.1	97.6
Sugar-yielding products	kg/person	76.5	132.9	73.7
Pork, beef and mutton	kg/person	27.5	42.8	55.6
Aquatic products	kg/person	14.9	17.2	15.1

Notes: Figures are derived by dividing the total output value or volumes by the total number of labourers in the countryside. Figures for output value are in current prices.
Source: *NCTJNJ 1985*, p171.

storage capabilities and transportation capacity. This creates another constraint on the development of specialized and market oriented agricultural production (Smil 1981, p76). It is one of the main reasons why inter-provincial grain trade was only 7.7 million tonnes on average for 1953–57 and 11.6 million tonnes in 1978 (Walker 1984, p184). In the early 1980s, higher levels of specialized production and a boom in output of the major agricultural products made the situation much more acute and the "two difficulties" (difficulty in selling agricultural products and buying production inputs and consumer goods) were the subject of complaints by peasant farmers almost everywhere (He 1984, pp2–3).

The market separation Perhaps the most significant constraint on the development of market orientated agriculture is the monopoly controlled market system imposed by the government. The market separation of supply and demand has meant that most peasant farmers had no choice but to grow grain for feeding themselves. The situation was more serious when regional self-sufficiency in grain production was stressed and free market trading prohibited.

5.2 Policy Mechanisms and their Impacts on Agricultural Production

Based on the critical policy issues identified and the two alternative policy patterns hypothesized in Chapter Two, which were validated in the context of output response estimation for grain products in Chapter Four, this section explores further the mechanisms of the policy patterns. This means examining closely the internal and external linkages of the agriculture sector, and their impacts on agricultural production.

5.2.1 Chinese agriculture: beneficiary of or sufferer from development policy?

Before approaching the internal linkage of policy issues and their impacts on agricultural development, it is necessary to identify the external linkage, that is to clarify the controversial topic of whether the agricultural sector has benefited or suffered from the intersectoral resource transfers. This is a precondition for understanding the nature of the policy patterns and their impact on agricultural production.

As mentioned in section 2.2.2, many Western economists believed that the Chinese development path was fundamentally different from that of the traditional Soviet model in that the intersectoral trade in China continually improved in favour of agricultural development. This argument is supported by three pieces of evidence. (1) The tax on agriculture was not increased in line with the growth in production, its proportion in the state budget income decreased with time (see Table 2.3). (2) Since the 1960s, the production of more industrial inputs for agricultural production was emphasized which resulted in a significant improvement in agricultural production conditions (Table 1.3). (3) The official price index between purchasing prices paid and the prices of industrial goods purchased by farmers suggests that Chinese agriculture has benefited considerably from an improvement in intersectoral terms of trade (Abbott 1977, p320; Lardy 1978, pp176–178). However, an alternative view suggests that China's agricultural sector was generally squeezed in the last three decades (Lardy 1983). Furthermore, it is suggested that the extent of the squeeze changed with time as policy shifted between Policy Pattern A and Policy Pattern B. The evidence for this view will now be considered.

First, it is argued that the official price indices concerning agriculture's terms of trade are not a reliable analytical tool. The calculation of the price index of industrial goods sold to the countryside is based on the prices of cotton cloth, refined sugar, refined salt, kerosene and matches and without any consideration of the prices for major industrial inputs, such as agricultural machinery, chemical fertilizers and sprays (see *ZCWTZ 1952–1983*, pp433–439 and *NCTJNJ 1985*, pp148–151). Moreover, since the trade condition between the agricultural and industrial sector was worse in the early 1950s than before the Second World War (see section 2.5.2), taking 1950 as the basic year is misleading. Using such indices to measure the real intersector trade is not helpful. The real issue, according to Chen (1983, p96), is that the price-gap between industrial and agricultural goods widened considerably during 1952–81. The original 22.6% undervaluation for agricultural produce in 1952 went up to more than 50% for 1959–73, and finally reduced to 30.3% in 1981, which is still higher than the base year of 1952.

Second, direct comparison between prices of major industrial inputs and agricultural products tells the same story. Taking domestic fertilizer production and supply as an example, while production costs of urea were only 150 yuan per tonne in the modern plant of Heilongjiang Province, the selling price to farmers was 450 yuan per tonne. Of this margin, 200 yuan was taken by the factory as the profit mark-up and 100 yuan was charged by the marketing system for distribution (Tong and

Bao 1978). As a result, the 5.26 ratio of nitrogen price to paddy rice price in China in 1980 is significantly higher than all other Asian countries considered (Booth *et al.* 1986, p28). The relative price between agricultural products and farm machinery is also seriously distorted. For example, in Japan a conventional 20-horsepower tractor sells for the equivalent of 5.5 tonnes of rice, whereas in China the corresponding Chinese model sells for 35.5 tonnes of rice (Chia 1979, p15). Although such a comparison is distorted by the protection policy for rice production pursued by the Japanese government, the general conclusion would not be changed even taking into account the inflated rice price in Japan. The Chinese model is still over twice the world price (Lardy 1983, p115).

A third indicator of the price squeeze on agriculture is provided by comparing prices of agricultural raw materials and the prices of the resulting industrially processed goods. By establishing itself as the monopolist purchaser of major cash crops and the monopolist seller of processed goods both in domestic and international markets, the government generated large profits. Taking cotton as an example, the price ratio of ginned cotton to grain products for most of the period from the mid 1950s to 1978 was lower than that in 1952 which itself was equal to the pre-war level. (Appendix Table 1.3). Also, the price government set for cotton cloth relative to ginned cotton in 1950–55 was 29% and 21% higher than in the pre-war period in Shandong and Sichuan Provinces respectively according to Li (1980, p50). Because its major input was underpriced and its products overpriced, the cotton textile industry has been one of the most profitable industrial sectors, about 10% of the state budgetary revenue was contributed by this sector in the late 1970s (Ministry of Textile Industry Research Office 1981, IV–48). In the case of the sugar refining industry, the production of one tonne of refined sugar produced 740 yuan profit and taxes for the factory in 1979, which is 2.36 times as much as the material cost (314 yuan per tonne) and 45% higher than the total production costs. The retail price of 1,540 yuan per tonne is 4.9 times as great as the value of its agricultural material input (Hsaio 1980, p47). For light industry as a whole, the manipulation of the prices of agricultural inputs and the selling prices of its final goods by the government made the sector the major source of the state budgetary income. During 1952–77, 29% of the whole state budgetary revenue was contributed by the light industrial enterprises although its share of the national income was only 13% in the 1950s and less than 16% in 1978 (Ishikawa 1965, pp66,76; Yang and Li 1980, p20). These funds were equivalent to 70% of the total state investment during the same time period, most of which was used to finance investment in heavy industry, and less than 8% of these profits were reinvested in light industry itself (Lardy 1983, pp126–127). Since it is Chinese peasant farmers who provided most of the raw materials and consumed about two thirds of the light industrial products, they are the real contributors. It is through these mechanisms that agriculture's financial contribution was switched from a direct agricultural tax, which fell from 29.3% of the state budgetary revenue in 1950 to 2.6% in 1984 (*TJNJ 1985*, p524), to the "indirect agricultural tax" policy which enabled the government to get significant resources from the industrial sector, especially from the "profits of light industry".

A fourth way of assessing agriculture's term of trade is to compare the official purchasing prices with those in the free market. Despite the large increase in purchasing prices for major agricultural products since 1979, the state procurement price is still well below the market price. According to Lardy (1983, pp120–121), in the first quarter of 1980, the quota prices for wheat and corn were only about half the average prices prevailing in the rural market and the over-quota prices were 20% lower than the market price. For rice, the rural market price was 2.3 times the state's quota price and more than 50% higher than the over-quota price. For the cereals as a whole, the composite average price paid by the government in 1980 was about half the price of the rural market. This argument is confirmed by the official statistical data presented in Table 5.4. As can be seen from the table, although the price ratios of major products between free market and official procurement fell somewhat between 1979 and 1983 (except for the aquatic products) due to the increase in purchasing price and the enlargement of free market trading, until 1983, market prices were significantly higher than the official purchasing price. The margin ranged from 2.3 times for cereals to 1.2 times for cotton, tobacco and hemp. Although state extractions of grain and edible oils from the countryside at well below the market price did not lead to large state profits directly, it is a most important factor explaining the low wages of urban employees, (Table 1.12) which stagnated for 27 years between 1957 and 1983 (*TJNJ 1985*, p556). At the same time the gross value of industrial output, state profits and tax revenues from the industrial trading enterprises and the labour productivity of the nationalized industrial sector increased 9.2, 5.4 and 1.1 times respectively (*TJNJ 1985*, pp25,382, 523). This provides further illustration of the squeeze on agriculture in support of the industrialization programme by the government.

In summary, on balance, the evidence suggests that, instead of having benefited from the course of development, agriculture has actually been substantially squeezed since the early 1950s. The major way of achieving this was through compulsory purchasing of grain and oil-bearing products at low prices. This enabled low labour wages in the industrial sector and by the creation of monopolized purchasing of

Table 5.4 Price indices of the free market for major agricultural products, 1979-1983 compared to official purchasing prices

	1979	1980	1981	1982	1983
Grain	241.9	230.5	234.7	237.2	230.0
Fat & oil-bearing products	235.4	217.7	207.3	191.6	188.5
Cotton, tobacco & hemp	136.0	130.0	137.8	127.3	123.7
Meat, poultry & eggs	142.5	122.4	126.9	131.0	132.2
Aquatic products	167.9	155.0	163.2	177.0	186.8

Note: The official purchasing price is the composite average price of quota, above-quota and negotiated purchasing offered bystate marketing system.
Source: *ZCWTZ 1952-1983*, p398.

other agricultural products (mainly industrial crops) and selling of consumer goods and industrial inputs the government was able to extract huge profits from industry and commerce, especially from the light industry sector.

5.2.2 The mechanism of policy patterns - its internal linkage

Following the theoretical hypothesis of the policy patterns and mechanism described in section 2.6 and bearing in mind the results derived in the last chapter and discussions above, it is time now to explore further the internal linkage of the policy issues, in order to understand better the logic behind the policy mechanism and how it worked in practice. For illustration, the discussion is once again focussed on the Policy Pattern B which relates to the periods of the Great Leap Forward and the Cultural Revolution.

Given the fact that China was primarily an agricultural society with a backward economy, the only way the government could launch an ambitious development programme which gives heavy industry sector the priority is to squeeze agriculture. This has to be achieved by means of intersectoral transfer of labour, capital and consumer goods. Since labour has always been plentiful in the nation, the major problems were to transfer a considerable amount of capital for new investment and to provide sufficient cheap consumer goods, which in China's case are grain and other major agricultural products. These are the ways to provide a cheap and sufficient labour force to build up the heavy industrial sector.

The direct way to extract capital and goods for basic needs and as raw materials from the agricultural sector is through the pricing and marketing system. First, the price for basic needs has to be cheap. This implies the price has to be directly controlled. However, the control of price itself is not a sufficient condition for guaranteeing the availability of goods. Thus centrally manipulated pricing has to be backed up by an interventionist marketing procedure, that is, the monopoly control of supply and demand. By this means, the intersectoral capital transfer (mainly from indirect tax) and provision of sufficient goods for basic needs are supposed to be achieved simultaneously. However, market control and price manipulation does not necessarily mean that producers' production decisions can be automatically steered onto the desired course. For example, when price for one product is too low and the production activity becomes unprofitable, producers may shift their production to another and some might even give up farming altogether. Consequently, a centralized direct controlling system has to be imposed in terms of production decision-making and management. In this system, the direct planning targets would tell producers what, how and how much to produce. This has in turn to be backed up by the market separation of supply and demand for agricultural products as well as industrial inputs. In this case, if a farmer did not grow grain crops, there would be nowhere for him to get grain for self subsistence. However, given the fact that China is such a huge country with hundreds of millions of peasant households, this necessitates setting up rural institutions and production organization. All the market, pricing and production control instruments could not otherwise be implemented and work out properly in practice. This is why agricultural collectivization with the main features of pooled labour and egalitarianism in income distribution became the fundamental issue which had to be solved simultaneously. This is the economic logic of how and why these three key policy ingredients have to be designed and implemented together.

The outcome of this theoretical reasoning is fully exemplified by the two periods of the Great Leap Forward (1958-60) and Cultural Revolution (1966-77). During these periods the large scale basic production organizations were set up. Labour and other production means were merged, labour was pooled for unique action and "everyone eating from the same public pot" practised. At the same time, a centralized management system was imposed and production plans drawn up by the central authority were passed down step-by-step to be directly implemented by the basic production units. Meanwhile, monopoly market control was strengthened and the price gap between agricultural and industrial goods widened. On the other hand, in order to secure grain production given such an unfavourable general climate, the price ratio between grain and cash crops was manipulated against cash crop production. Furthermore, free market trading was prohibited, peasant households' sidelines were constrained and high purchasing ratio for grain products was enforced given the low level of per capita production. Unfortunately, all these measures restricted the peasant households' role in production decisions and management and they paid little attention to the disincentive effects on production. In practice, the Policy Pattern B did not work out well and indeed created considerable damage to the development of China's agriculture as well as her industrial sector.

5.2.3 Policy impact analysis

Under the Policy Pattern B in which collectivization, centralized management and monopolized pricing were the main policy measures, the peasant farmer's initiative and enthusiasm for agricultural production was dampened, great economic losses thus resulted from poor management and acted against comparative advantage. The major economic consequences from executing such a policy pattern are discussed below.

Collapse or stagnation in agricultural production A serious crisis in agricultural production was created following the Great Leap Forward and People's Commune movement. According to the official statistics, the gross value of agricultural output in 1961 was 5.9% less than in 1952 and 30% less than 1957 (in 1952 price) (*TJNJ 1983*, p149). The reductions in output of the major agricultural products in 1961 compared to 1952 were 2.42% for grain, 42.5% for cotton, 52.2% for oil-bearing products, 50.2% for sugar-yielding products and 42.7% for meat (pork, beef and mutton) and the corresponding reductions were 18%, 54.4%, 52.3%, 68.2% and 51.3% respectively compared to 1957 (Table 1.5). Generally speaking, the Great Leap Forward and People's Commune movement set back the development of Chinese agriculture for more than ten years.

During the period of Cultural Revolution, although the total value of agricultural output grew 3.2% per annum which was partly due to the development of rural industry (*TJNJ 1983*, p149), the 2.6% annual increase in grain production during 1966-77 only just covered the 2.5% annual growth rate of rural population for 1965-77 (*TJNJ 1983*, pp103,158). For the oil-bearing and sugar-yielding crops, the 0.8% and 2.3% output growth per annum during 1965-77 actually reduced the per capita level of output. During the same period, cotton production registered a net reduction of 12.3% in output (*TJNJ 1983*, pp159-160).

Poor economic efficiency When Policy Pattern B was pursued 'political accounting' was far more important than 'economic accounting'. In practice, there was little room for consideration of economic efficiency in production management and decision-making, although collectives were required to balance their income and expenditure. During the era of People's Commune movement, the mistakes in decision-making and poor management wasted a great deal of human energy as well as other economic resources. Indeed, given that production was set back for more than ten years, there is nothing that can be justified in terms of economic efficiency. For the period of the Cultural Revolution, agricultural productivity fell substantially. According to figures compiled by Tang and Stone (1980, p28) the aggregate input index for agricultural production increased 63.8% (from 141 to 231 based on an index of 1952 = 100) during 1965-77, while the index for gross value of agricultural output grew only 40% for the same period (from 147-197). This was confirmed by Du (1984, p30) who claimed that the labour productivity in 1978 was lower than that in 1955.

Regional economic efficiency can be checked from the encouragement or discouragement of specialized agricultural production. This is another clear-cut policy issue shown in the changes in policy pattern. When "taking the grain as the link" was pursued as the strategy for agricultural development, regional food grain self-sufficiency was stressed and this worked against the specialized and market orientated production creating enormous losses in terms of allocative inefficiency within agriculture. This issue has been systematically investigated by Lardy (1983, pp48-87). Typical examples are the trade-off between grain and animal husbandry development in the Xinjiang autonomous region; grain and cotton production in Shandong and Henan Provinces; grain and sugarcane in Fujian Province; and grain and peanut production in Shandong Province. The general lesson from all these examples is that when "taking the grain as the link" was stressed and when regional self-sufficiency targets for grain were set, not only was the production of cash crops and the livestock industry set back badly, but worse still grain production itself also stagnated or decreased. However, then diversified development and specialized production according to the comparative advantage was encouraged, the production of grain and other sectors boomed simultaneously.

Increased agricultural imports Due to the drastic output reduction in domestic grain production caused by the People's Commune movement, China was forced to change her position from a net grain exporter in the 1950s to the net importer in the early 1960s. This was continued throughout the Cultural Revolution era despite attempts by the government to achieve the goal of self-sufficiency (see section 1.5.2). From 1960 to 1977, the net grain import amounted to 49.3 million tonnes (Table 7.2). Meanwhile, imports of other major agricultural products also shot up during the same period. The imports of ginned cotton, sugar and edible oils increased from 93,800, 234,000 and 490 tonnes in 1960 to 181,100, 1,598,000 and 137,229 tonnes in 1977 respectively. The 1.5 billion U.S. dollars for the import of grain, cotton, sugar and edible oils accounted for 21.3% of total imports in 1977 (*ZCWTZ 1952-1983*, pp511,514).

Effects on farmers' incomes and provision of basic needs Due to the reduction or stagnation in agricultural production as well as poor economic efficiency, producers' income and the provision of basic needs were seriously affected. For example, per capita net cash income of the rural population fell 14.4% between 1957-62, and achieved only an average growth rate of 3.1% during 1966-77 (in 1952 prices) (*ZCWTZ 1952-1983*, pp20, 370). In terms of providing basic needs, the per capita consumption of grain, edible oils and pork in rural areas registered 20.1%, 57.3% and 59.1% net reductions respectively during 1958-62, and with complete stagnation for 11 years from 1966 to 1977 (*ZCWTZ 1952-1983*, pp27-29). As a result a net reduction in rural population was recorded during 1958-60 with -9.23% natural growth rate in 1960 (*TJNJ 1983*, pp103, 105) and more than 10 million of the rural population were still suffering from insufficient food grain until the end of the 1970s (section 2.5.2).

Effects on the development of the industrial sector Since China is a country under transition from an agricultural society to an industrial society, the development of agriculture is the precondition for the development of industrial sector as well as the whole national economy. The set-back of agricultural production at the end of the 1950s and stagnation during the Cultural Revolution period imposed an enormous negative impact on the development of the nation's industry. This relationship is clearly shown in Chart 1.1.

5.3 Conclusions and Lessons

5.3.1 Conclusions

The policy measures and their impacts on China's development when Policy Pattern B was pursued is summarized in Figure 5.1. Bringing these conclusions together with the discussion in Chapter Two and the results of Chapter Four it is now possible to supply answers to two of the initial questions posed in the Introduction: viz. that Chinese peasant farmers are economically rational producers, both in history and at present time. Like most farmers, China's agricultural producers responded to the price signals quickly, normally and efficiently. For the last three decades since the founding of the PRC, no matter how the policy changed from time to time, the key elements determining grain producer's supply response behaviour remained household self-subsistence and profitability. Achieving self-subsistence in China is not just determined by the conventional constraints of technology and social infrastructure, these are significantly compounded by the compulsory purchasing system imposed and market separation between supply and demand. The material effect is therefore that peasant farmers faced more constraints in pursuing the aim of economic rationality: it does not necessarily mean that producer's initiative and

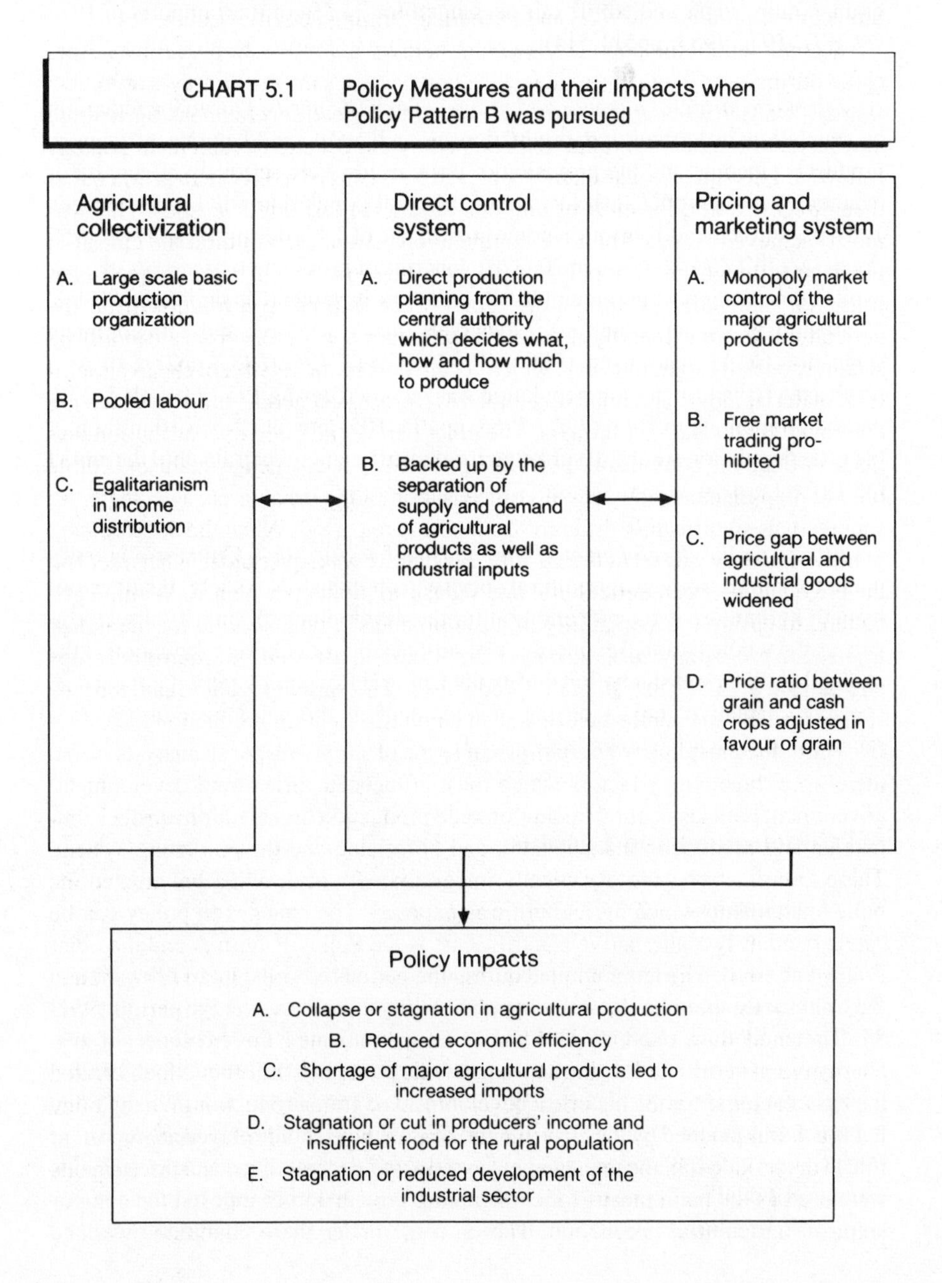

enthusiasm for profit is eliminated. The reason is very straightforward, for even under the socialist setting of the Chinese economic system, agriculture has never been nationalized. Chinese peasant farmers and basic production units have always had to manage to make ends meet financially. Their daily lives are not guaranteed in any effective way by any authority. Given the very low level of income and low provision of basic needs, peasant farmers have to fight in every way they can to protect their families' survival and keep their farming activities going.

Second, unlike the popular viewpoint held by many Western scientists especially during the 1970s, this evidence weighed above confirms Lardy's argument (Lardy, 1983) that under the socialist setting of Chinese development, agriculture as a sector was squeezed rather than helped by the general development strategy followed. This squeeze was mainly achieved by means of indirect extraction rather than direct taxation for most of the time during the last three decades. This was mostly achieved by means of price manipulation between agricultural and industrial goods and through the monopoly rôle played by the government as the unique buyer, processor and seller of agricultural goods. The indirect contribution from the agricultural sector was partly channelled by the monopoly profit from the industrial sector, especially light industry, which resulted from the relatively cheap agricultural material inputs and relatively expensive processed goods, most of which are consumed by the peasant farmers. The other part of the indirect contribution was made through the compulsory purchasing of basic goods (such as grain and edible oils) to help maintain low labour compensation in the urban area. The degree of squeeze was significantly different from period to period. When the development of heavy industry was set as the priority, agriculture was squeezed much harder and the price gap between agricultural and industrial products widened. Within agriculture, to ensure the availability of grain products which are vital for the industrialization programme to be sustained, the price ratio between grain and other crops was adjusted in favour of grain production. This pricing policy can thus be summarized as a general squeeze and an emphasis on grain production.

Third, the most important findings in terms of supply response analysis is that there were three policy factors which most influenced agricultural development: government policies on rural institutions and production organization; on decision-making and production management; and on pricing and the marketing system. These three factors were organically integrated with each other and created the policy patterns to which agriculture was exposed. The changes in policy can be categorized as two alternative patterns. One is the Policy Pattern A and the other Policy Pattern B. The latter applied during the periods of Great Leap Forward and the Cultural Revolution, the former applied at all other times over the period 1952-84. The modelling exercise in Chapter Four confirmed the existence of two alternative patterns and it is the shifts from one pattern to the other which created the zig-zag course for the historical development of grain production. When Policy Pattern A was pursued by the government, peasant households played an important rôle in the production, the management system was decentralized and price signals were used as the main means for encouraging growth and changes in the composition of agricultural production. The outcome under these conditions was the successful development of agricultural production and a significant improvement in peasants' daily lives. The main external effect was the sustained development of the industrial sector as well as the whole economy. However, when the development of heavy industry was stressed and Policy Pattern B imposed, individual producers' responsibility was submerged, the management system was centralized, purchasing price stagnated and more strict control of the marketing system was pursued. Under these conditions, production crashed or stagnated, farmers' living standards worsened and consequently the development of industry and the whole economy was set back.

5.3.2 Lessons from the historical development

Several important lessons can be learned from the historical agricultural development from the policy point of view. One is the development of Chinese agriculture is critically dependent upon the attitude and responsibility of the hundreds of millions of peasant farmers. Sound understanding of the producer's supply response behaviour with appropriate protection of their interest to encourage their initiative and enthusiasm for farming is the fundamental condition for the successful development of the agricultural sector.

The second is that given the nation's huge farming area and complicated farming system with its great regional differentiation, the production management and decision-making systems must retain great flexibility allowing local conditions to be taken into account. The policy indication is that centralized management has to be integrated with decentralized management, the co-operative ownership and its responsibility for decision-making and production management has to be protected. Moreover, policy measures for encouraging the development of specialized production should be pursued to make better use of the natural and economic resources.

The third lesson concerns the trade-off between agricultural and industrial development. For the transition from agricultural society to the industrial society, the primary accumulation of capital and other resources has to be largely contributed by the agricultural sector especially in the initial stage. This is a common phenomenon all over the world no matter which social system is chosen (Chen 1983a, p91). However, given the fact that progress in agriculture is a prerequisite for industrial development, which is certainly the case in China, a balanced development which can promote overall development and structural transformation and take full advantage of positive interactions between agriculture and other sectors is the appropriate strategy to pursue. Policy Pattern A is therefore the preferred way for achieving the aim of China's industrialization. It is against the background of the extremely heavy price paid by Chinese peasants for misguided agricultural policies that systematic policy changes have been taking place since 1978. These provide the hope for the future of the nation's agriculture.

Having focussed thus far on supply response analysis, the next chapter switches attention to demand concentrating on factors affecting domestic grain consumption.

Chapter Six

Domestic Consumption of Grain Products

6.1 Introduction

The objective of this chapter is to identify the critical factors which influence the domestic consumption of grain products. The approach is to use elementary regression techniques to estimate consumption relationships based on aggregate time-series data. The estimation of consumption is disaggregated into four categories; food, which is the most important part, feed, seed and industrial usage. In view of the distinctive distributional systems existing between urban and rural areas in the last three decades, food consumption is further divided into that relating to urban and rural areas. The broad theoretical hypothesis to be tested is the relationships between food grain consumption and consumers' income level, price signals, the production level available and the growth in population. The chapter starts with a general review of the grain distributional system.

6.2 Review of the Grain Distributional System

Mao Zedong once said, "Planned purchases and planned marketing are important steps towards implementing socialism" (*People's Daily* 1977, p2). The scheme of Central Purchase and Supply has been implemented by the Chinese government for more than three decades. While the purchasing aspect has been reviewed in previous chapters, the grain rationing system, which is vital for domestic consumption should be described before specifying the demand functions.

Grain, as well as other food products, has been rationed since November 1953 (Smil 1981, p72). On August 25th, 1955, two files concerning the "temporary measures for rationed supply of grain for city and township people" and the "temporary measures for planned purchases and marketing of grain" were issued by the State Council (*People's Handbook* 1956, p488). It was at this time that the planned purchasing and marketing for urban and rural areas became formalized (Chen 1978, p29; Walker 1984, p68). Under the planned marketing system of food grain products, citizens in city and township are divided into nine categories by the administrative system according to the age and employment status of each person: unusually hard physical labourers, hard physical labourers, light physical labourers, white collar employees, college and high school students, general public and children over ten, children between six and ten, children between three and six and children under three. The monthly standard supplies for each group and the actual national supply in the 1950s has been discussed by Chen (1978, p31). Urban citizens purchased grain with grain tickets and money at designated shops. The actual ration levels, the scope of rationing and the composition of grain supplied (the proportion of fine grain) changed from time to time and with considerable regional differentiations (Walker 1984, pp67–76).

Planned purchases and marketing of grains in rural areas was initially established by using the rural household as the basic unit. However, due to the transitions in rural institutions since the late 1950s, the production team was designated to take its place until the implementation of the household responsibility system at the end of the 1970s. So for a considerable time, production teams were responsible for grain production, marketing and internal consumption. After submitting a certain amount of grain for tax and quota purchase, the production team has to allocate grains for seed, feed, sideline occupation and grain reserves. The remaining grain is then allocated to its member families for self-subsistence. Generally speaking, 30% of this remainder grain is distributed as "work-point grain" according to the work points earned by each peasant household, and the 70% as "basic grain" which is to be distributed to each person. For those production teams which do not produce grain or are not mainly engaged in grain growing, they can be exempted from supplying grain under compulsory purchases with permission from the administrative authority, and supplied with food grains in the same way as in cities and townships (Chen 1978, p30). On occasions when basic production units suffer from natural calamities, compulsory purchase quotas may be exempted for the current year and "relief grain" may be issued by the social security department. With the contracted responsibility system executed since 1978, peasant households became the basic units for grain production and distribution. After submitting fixed amounts of grain products as the obligations to the state and the local government according to the contract made with the co-operatives, peasant farmers have the full responsibility for allocating the remaining grains for different purposes.

The general principle used in implementing the grain rationing scheme is "the combination of unique balancing and multi-stage management". With this principle, annual meetings for central purchase and supply of grain is held by the State Council which decides the general equilibrium of grain production and consumption. The general balancing account drawn by the Ministry of Grains specifies the total purchasing, total supply for domestic consumption (including urban food grain supply and rural grain resales), the adjustment in national grain reserve and the level of imports and exports. It also spells out the detailed figures for provincial purchasing, supplying and trade from one province to another which is then implemented by provincial governments. Grain consumption of the metropolitan cities of Beijing, Tianjin and Shanghai, the army, the state-owned industries, the relief grain and foreign trade are directly controlled by the central government (Wiong 1980, p125). Details about the marketing institutions under the central purchasing and supply system can be seen in Abbott (1977, pp318–330).

6.3 Model Specifications and Results

Analysis of the domestic consumption for grain products in this study is disaggregated into four categories, (i) food grain for human consumption, (ii) feed grain for livestock production, (iii) seed grain for crop cultivation, and (iv) the industrial demand of grain products. The first category is a primary demand and other outlets are of a derived demand type. All are analysed at the national aggregate level. Bearing in mind the fundamental differences in the distributional system in town and country as reviewed above, consumption of food grain is further divided into urban and rural aspects. However, mainly due to the problem of data availability, no further disaggregation is made into the detailed composition of demand, such as the detailed categories of food grain, the feed grain demand of individual classes of livestock production or the variety (or class) of seed grain for individual crops. Furthermore, although the national reserve is another important source of domestic grain demand in China, time-series data on stock holdings has never been published for external usage: it is therefore omitted.

6.3.1 Food consumption - model specification

Conventionally, the basic unit of demand theory of neo-classical economics is the individual consumer or household (Tomek and Robinson 1981, p25). With the operation of market system, a consumer selects the specific goods and services that give him the highest satisfaction. In short, the neoclassical view of consumer behaviour is that he selects goods which maximizes his utility, given the market prices and his budget constraint. Therefore, a consumer's demand for goods depends on the prices, his income and preferences (Thomas ed. 1972, p18). For the aggregate market demand of goods, the major factors influencing the level of demand have been grouped as, (1) population size and its distribution; (2) consumer income and its distribution; (3) prices and availability of other commodities and services; and (4) consumer tastes and preferences (Tomek and Robinson 1981, p32).

In demand theory, the change in price is taken as the "cause" and the change in quantity of demand as the "effect". For various reasons, the effect is likely to be spread through time instead of occurring instantaneously at a point in time (Tomek and Robinson 1981, pp40–41). In this case, a distinction between short-run and long-run demand has to be made in coping with the delayed adjustments associated with the passage of time. Like that of estimating supply relations in agriculture, distributed lag mechanisms can also be employed for taking account of the effect of delayed response in demand relations.

Another critical aspect to be considered in empirical modelling of food supply-demand relations is whether there are two way causation relationships. When there exists two way causation between supply and demand, the application of ordinary least squares to a single demand equation would violate the assumption that the expected value of the error term is zero and consequently yield biased and inconsistent estimates (Koutsoyiannis 1984, pp331–335). Under a market economic system with the assumption of competitive price determination, a simultaneous equations model is usually built which determines price and quantity together (see Tomek and Robinson 1981, pp303–306; Koutsoyiannis 1984, pp331–332).

For the specification of food grain consumption functions under Chinese circumstances, key factors which dictate the characteristics of consumption patterns must be clarified. First, there has been very low per capita availability of grain products during the last three decades. The majority of the rural population are peasant households with a predominantly self-subsistence orientation. Therefore, the consumption of food grains in China is vitally dependent upon the production level achieved. This feature is a very important restriction on the utility maximizing theory of consumer behaviour summarized above. Quite clearly, in Chinese circumstances where output is often deficient in the absolute sense of providing less grain than is desired and occasionally less grain than required for minimal subsistence, consumers cannot be viewed as optimally allocating their budget amongst foodstuffs in response to changes in relative prices. Their ability to manipulate purchasing is often limited by availability, and prices themselves do not always reflect accurately the scarcity of grain because of the rigidities of the political controls exercised. Thus, the analysis reported in this chapter cannot strictly be called demand analysis with its connotation of consumer choice restricted only by the budget constraint. Instead the approach is to investigate variables which explain the variations in grain consumption observed in China. More detailed arguments on this matter are presented when discussing the independent variable of grain output below.

Second, to launch the industrialization programme given low level of grain production, the market for grain products was monopolized by the government and supply and demand were thus isolated. The combined effect of an imposed rationing system and constrained free market trading for most of the time rendered price and income largely ineffective and a predominant rôle in determining the level of food grain consumption was played by production and population growth. Third, the strict control of consumption and market separation prevents feedback from demand to supply particularly while production determines the consumption level. Fourth, with the rationing system imposed, the price for food grain products is fixed by the administrative system and promulgated by the official marketing agencies. So, just as in the supply response analysis in Chapter Four, there is no need to introduce a dynamic adjustment mechanism in the specification of demand functions. Finally, given that the consumption of other major consumer goods was also directly controlled by the government with the quantity of supply rationed and price level fixed for most of the time during the last three decades, there was therefore no significant competition between the consumption of grain and other major consumer products. All these characteristics imply that the consumption function for China's food grains is aggregate, static and recursive in nature. A recursive econometric model is appealing in this circumstance, it is appropriate to use OLS method to estimate the parameters of behavioural equations involved. The theoretical consumption function in this research is therefore specified by relating the quantity of food grain consumed to grain output in the same or previous period, population, disposable personal income and the price of food grain products. The definition and specification of each of these variables are described below and this is followed by the results of the estimation of consumption equations for each category of grain.

The dependent variable: quantity of food grain consumed Four major issues deserve clarification about the food grain consumption in China. The first is the confusion about its statistical definition. Interpretation of these statistics is regarded as puzzling for many foreign researchers (Walker 1982, pp575–588). The current statistical system in China defines grain production figures, (output and yield) measured in unmilled form. These are the figures collected by the Agricultural Ministry (also by the Bureau of Agriculture, State Statistical Bureau) and have been used in the supply response analysis of Chapter Four. However, trade grain (or milled grain in this study) is the item used for grain consumption and trade statistics by the grain administrative, commerce and trade management departments. Thus trade grain is used in the analysis of domestic consumption and foreign trade in this chapter. According to the information from the Bureau of Commerce, State Planning Commission of China, the conversion ratio from unmilled to trade grain is 65% for rice, 75% for millet and 100% for others. The general conversion rate for grain as a whole is 85%, which is the figure used by Li (1959) and Walker (1982).

The second concerns the difference in statistical accuracy between urban and rural consumption. As reviewed in section 6.2, domestic grain consumption has been under strict control for the last three decades and the differentiation between the urban and rural distributional system is fundamental. For urban grain consumption, a clear norm is set and food grain is allocated according to the number of citizens, their age structure and employment status. The statistics derived from such a stringent system leave little room for any kind of misunderstanding. However, since most of the peasant households and basic productions units are grain producers (wholly or partly), and the performance of grain distribution is based on the basic production unit for most of the period analysed there is greater scope for errors in records of consumption. There are several ways in which retentions of grain for the peasant household's self-subsistence may be disguised. For example, the yield and output could be underestimated, and more grain for feeding animals, seeding and the collective reserve might be declared. Since the livestock industry is mostly small scale household sidelines (pork makes up 95% of the domestic meat supply but pig raising is mainly a household sideline), grain in a peasant family is usually consumed by humans and its remainder or by-products from processing is used for feeding animals. All these factors demonstrate a much more flexible grain consumption pattern in rural areas compared to the urban areas. These are mainly caused by the combination of low grain consumption and low income level prevailing in the vast Chinese rural areas. From the Chinese author's experience, peasant farmers in his home town had to save some grain every year to sell to the free market to serve their families' basic needs and remain in production. Although it is not clear what impacts these considerations have on the data for rural consumption, there must be a general expectation of systematic downward bias in rural consumption data, especially during the periods when peasant households had more discretion over rural grain distribution.

The third feature of grain consumption concerns the very different grain consumption conditions between urban and rural sectors, both in quantity and quality. Generally speaking, grain supply for urban consumption is much more secure and stabilized than that for rural areas (Walker 1984, pp68–70). From 1952 to 1983, per capita consumption of milled grain in city and town fluctuated between 179.5 kg (1961) and 242 kg (1953) while that for rural areas was between 154 kg (1961) and 235 kg (1983) (Appendix Table 1.6). Due to the greater quantities of grain supplied and the additional food consumed by urban citizens (see Table 1.14), the majority of the urban households had surplus grain stored in the designated shops. In contrast, for most of the peasant families especially those in southwest and northwest areas, shortages of grain for family consumption prevailed for most of the last three decades. According to the Chinese author's experience in Zhiyang County of Shanxi Province (northwest China) in the early 1970s, the grain which peasant households obtained from the production team was enough for three months' consumption only, for the other nine months the rural population had to rely on government "relief grain" which was on average 5 kg per month for each person. Malnutrition was a common feature in that area during that time.

There were also obvious differences in the composition of grain consumed. The general picture is better quality of urban consumption with high proportion of fine grain (rice, wheat and soybean). For most people in rural areas, sweet potato constituted the main food for the majority of the peasant families in the south, and maize and sorghum were the main energy sources for those in the north (Walker 1984, Ch 2).

The fourth main feature of the nation's grain consumption concerns the existence of significant geographical differentials which stem from the basic self-subsistence nature of China's rural areas. This is illustrated in Table 6.1. In 1980, the average per capita grain consumption of the nation's rural population was 257 kg, of which 163 kg were of fine grain. However, the per capita consumption level in Zhejiang, Jiangxi and Hunan Provinces were respectively 21%, 19% and 30% higher than the national average. In contrast, rural populations in Hebei, Shandong, Henan, Guizhou, Yunnan, Shanxi, Ningxia and Xinjiang Provinces consumed less than 90% of the national average level. The lowest level of average consumption was 206 kg in Shanxi Province. This amounted to only 80% of the national average. In terms of fine grain, Zhejiang, Jiangxi and Hunan Provinces enjoyed consumption levels 67%, 82% and 87% higher than the national average. However, consumption in Shanxi, Liaoning, Jilin and Heilongjiang Provinces are all under 40% of the national level. The worst region is again that of Shanxi Province. The 48 kg per head of fine grain consumption in this province in 1980 represents less than 30% of the average level of the nation.

The independent variables

1. *Grain output* Historically, consumption of food grain in China has been directly constrained by the production level achieved. As shown in Table 6.2, when

Table 6.1 The regional differences of rural populations per capita grain consumption (1980)

	Total consumption (kg)	In which: Fine grain (kg)		Total consumption (kg)	In which: Fine grain (kg)
Average of the nation	257.16	162.92			
Beijing	250.12	134.90	Jiangxi	306.41	296.48
Tianjin	238.42	151.12	Shandong	229.37	80.38
Hebei	226.22	68.21	Henan	221.60	123.03
Shanxi	241.93	48.36	Hubei	269.96	226.98
Inner Mongolia	253.07	75.34	Hunan	334.36	304.29
Liaoning	259.71	50.60	Guangdong	291.99	266.91
Jilin	299.20	64.86	Guangxi	257.17	234.71
Heilongjiang	253.88	58.44	Sichuan	258.52	184.03
Shanghai	296.92	238.51	Guizhou	224.67	149.19
Jiangsu	295.27	242.64	Yunnan	221.66	131.72
Zhejiang	312.44	271.84	Shannxi	205.89	90.50
Anhui	278.72	229.41	Gansu	238.34	120.09
Fujian	291.92	265.74	Ningxia	225.86	180.46
			Xinjiang	211.48	126.16

Notes: All the figures presented are for unmilled grain.
Data for Qinghai Province and Tibet autonomous region are not available.
Source: *NCTJNJ 1987*, p196.

Table 6.2 The trends of grain output and per capita food grain consumption

Year	Changes in grain output (unmilled)		Changes in per capita food grain consumption (milled)	
	Grain output (10,000 tonnes	Increase in %	Per capita food grain consumption (kg)	Increase in %
1952	16,392		197.7	
1956	19,275	+17.6	204.3	+3.4
1961	14,750	–23.5	158.8	–22.3
1965	19,453	+31.9	182.8	+15.1
1978	30,477	+56.7	195.5	+6.6
1983	38,728	+27.1	232.2	+19.0

Source: Derived from Appendix Tables 1.1 and 1.6.

production recovered from 1952 to 1956, per capita food grain consumption improved; when production collapsed between 1956 and 1961 with a reduction in output of 23.5%, consumption decreased 22.3%; when grain production increased from 305 to 387 million tonnes by a 27.1% total growth in output during 1978–83, per capita consumption improved 19%. While the fluctuation in production is due to the shifts in development policy as mentioned in previous chapters, the changes in food grain consumption are directly constrained by the shifts in production. Two points can be made in interpreting this direct correlation.

One is the nature of the low level of grain availability. Although the total grain output of 387 million tonnes in 1983 made China the largest grain producer in the world, per capita grain production of 378 kg is still below the world average of 400 kg (unmilled grain). Meanwhile, despite the fact that China increased her grain imports from world markets since the 1960s, the 11.47 million tonnes of net import in 1983 made up only 3% of the domestic production (see Appendix Table 1.1 and 1.15), which is thus not very significant in improving the domestic supply for the whole nation. Consequently, China's grain consumption is mainly influenced by the availability of domestic supply.

The second concerns the characteristics of the distribution pattern of grain products. The combination of the size of the country, the high proportion (76.5%) of the total population that lives in the countryside (1983), the lack of social infrastructure in the vast rural areas in the form of poor transportation facilities and marketing services, makes large scale grain redistribution impossible even if grain resources were available. So, it remains the case that most of the population in rural areas must be fed from their own farming activities. This self-subsistence therefore implies that output is one of the key factors in determining the consumption in any given year.

2. *Population growth* This is another critical element affecting the food consumption conditions. While grain production of the nation increased from 163.9 to 387.3 million tonnes, with a net growth in total of 1.36 times and 2.8% per annum during 1952–83, the total population for the same period grew from 575 to 1025 million, a 78.3% net increase. As a result, the 93 kg improvement of per capita grain output during the whole period of 1952–83 demonstrates only a 0.9% annual improvement (Appendix Tables 1.1 and 1.8). Turning to the consumption of food grain products, the 232.2 kg consumed in 1982 is only 17.5% higher than that in 1952, this represents an improvement of merely 0.5% per annum. The trends in grain output, population growth and per capita food grain consumption are shown in Chart 1.5.

Statistically, time-series data on population from the SSB is for the total number at the end of the year and further divided into the population in cities and towns and that in the countryside. This is not such a clear distinction as it, at first, appears, for there are peasant farmers living in suburbs and other workers with assured supplies of grain working in the countryside. Thus the population statistics are not totally identical to the two distributional systems as reviewed in section 6.2. Nevertheless, since the population figure is consistent with the per capita food grain consumption and other relevant statistics used in this study, it is appropriate for use in the analysis.

Like the explanatory variable of grain output, a positive relation between population growth and food grain consumption is expected.

3. *Personal disposable income* The inclusion of personal disposable income per head stems from the standard neo-classical theory of the consumer as a budget constrained utility maximizer. However, due to the differences existing in the payment system between China and Western countries as well as between urban and rural sectors of the nation, a general review of the wage system and its statistical implication is worthwhile.

(i) Urban wage system As there has been no private ownership of capital in the nation since the mid 1950s, wage earning has been the unique source of personal disposable income for most urban citizens. Starting in 1956, a Soviet type of payment system was adopted by the Chinese government. In this, most of the workers and employees in nationalized sectors were grouped into eight wage categories with a ratio between the minimum and maximum pay of approximately 1:3. In certain trades a larger number of ranks (up to 30) and a wider spread (up to 1:18) were introduced (Klatt 1983, p20).

During the first Five-Year Plan period (1953–57), wages for urban employees were improved slightly. However, for two decades between 1957–76, there was no improvement in wages at all, although labour productivity improved at the rate of about 3% a year and there was a marginal increase in the prices of consumer goods and services. The modest improvement in living standards of this period was achieved by the increase in number of employees of each family and the improvement in public services in urban areas. The average number of dependents per employee was 2.3 in 1957. This reduced to 1.06 in 1978, and 0.83 in 1980 (Lardy 1983, p164). A wage reform which emphasized higher payment for more and better work was introduced in 1978. The piece-rate wage system and system of premium payments for overtime work were also extended (Chen 1978, p21). Through these changes, together with the several-fold increase in wages offered by the government, the nominal wage payment per employee of the nationalized sector was increased from 602 yuan in 1977 to 865 yuan in 1983, an improvement of 43.7%. The real wage payment index increased 24% during the same period after taking inflation into account (take 1952 as 100) (*TJNJ 1985*, p556).

(ii) Rural payment system Unlike the payment system practised in the urban sector, the personal income of peasant farmers in rural areas has never been secured or directly controlled by the government. Their incomes come mainly from the selling of agricultural products and the offer of their labouring services. Thus the level of products produced and sold and the purchasing price offered by the government are fundamental in deciding the level of income they can achieve. When collective agriculture prevailed, the main income source for peasant farmers was the collectively distributed income according to the labouring contribution. Under such a collective income distribution system, rewards are calculated in three stages. First, labour was graded as adult male, adult female, and so on. The cultivator's initial remuneration is thus based on the work-points earned in the course of the year, for example ten work-points for an adult male per labouring day

and perhaps seven for an adult female. After deducting the production cost incurred and agricultural tax and other contributional duties to the central and local governments, total profit is calculated for the basic accounting unit. This mainly comprises the products distributed to the units' member families (calculated at the official purchasing price) and income earned from selling products (mainly the compulsory purchasing by the state) plus services to the outside. Third, the total profit made is then divided by the total work-points awarded to derive the real value per work-point. For those households which earned more work-points, that is more income in value terms than products they got from the basic production units, the remainder would be paid in cash at the end of the year. Logically, for those families which did not earn enough work-points to cover the products they received, cash had to be paid into the collective also at the end of the year. In general, payment in kind, for example in the form of grain, cooking oil and other agricultural products, accounted for about two thirds of the total collectively distributed income for the nation as a whole (Chen 1978, p29). During the era of the early 1950s and since the implementation of households responsibility system, the peasant family is the basic unit of accounting. After paying tax to the government and other duties to the local community, a peasant farmer has the full duty for financial management, both for production and family consumption.

Due to the continuing changes in rural institutions and production organization and the nature of highly dispersed and huge numbers of peasant households, an accurate measurement of the real income of the rural population is almost impossible. Currently, the indicator frequently used by both Chinese and overseas researchers is the "collectively distributed income", which is the remuneration received from collectives (see Lardy 1983, Table 4.6). Unfortunately, there are three main weaknesses which vitiate the value of this data to be used in this analysis. The first is the problem of incomplete coverage of income sources. Since it covers only the collectively distributed net income derived from the basic accounting unit, income earned by peasant households from selling produce of the private plots, from sideline productions and labour engaged outside the collective sector, as well as payments to the exchequer into welfare and so on are all excluded (Chen 1978, p79). This "non-collectively distributed income" grew much faster than the collectively distributed income in the last three decades, especially for the periods outside the Great Leap Forward and Cultural Revolution. All these render the collectively distributed income uninformative. Second, a large share of the collectively distributed income comes from grain and other agricultural products distributed in kind which are valued at quota purchasing prices as mentioned above. Thus an increase in price can substantially raise the income level without any change in the real goods distributed. As a result, highly inflated current prices made the collectively distributed income in 1961–62 significantly higher than in 1956–57 despite the fact that real agricultural production dropped substantially (Lardy 1983, pp160–161). According to Zhan (1980), the per capita collectively distributed incomes in 1977 (in 1957 price) of the nation was only 17% higher than 1957 instead of 60% displayed by the published statistics. In other words, more than two thirds of the claimed increase represents increased prices rather than real income. Finally, time-series data on collectively distributed income for 1952 to 1983 is not available.

Recently, information on per capita net income of the peasant households (see Table 1.13) which does not only involve peasant net income from the collective, but also from sideline production and employment on public works or in townships enterprises (previously the People's Communes) has become available. However, since this material was derived from sample surveys which took from only one in every 10,000 families and covered one quarter of rural counties in all provinces except Tibet, sampling and aggregation errors are inevitable. An upper bias is generally recognized by most of the Chinese researchers and practitioners for various reasons. Moreover, annual time-series data since the 1950s is not available. Consequently, despite its full coverage of the income made by peasant households, it was not considered appropriate for this analysis.

Net cash income per capita for the rural population has also been published (see Table 1.12). This time-series data is compiled by the Trade and Pricing Statistical Bureau of the SSB. The main advantage of this data is that it covers the whole range of possible cash income sources. The major weakness is it does not include the income distributed in kind, which is one of the major components of the peasant families' total income. Nevertheless, since it is the only time-series indicator available for the rural household's income, it is used in the estimation of demand.

Another indicator similar to disposable income is the statistic of per capita expenditure on consumer goods. This makes up the bulk of the citizen's cash income expenditure for both urban and rural areas in the last three decades and shows the same trend as the disposable income. It was therefore used as an alternative indicator for measuring the impact of income growth on food grain consumption. Naturally, a positive relationship between disposable income (or cash expenditure on consumer goods) and food grain consumption is expected.

4. *Price of food grain products* For most of the last three decades, the retail price for food grain supply in urban areas was cheap and stable. The retail index (based on 1952 = 100) for rationed grain price increased just 7% during 1952–60, another 15.2% in the next decade, and was then almost unchanged between 1970 and 1983 (Appendix Table 1.12). Before 1979, the cheap and stabilized price was maintained by the implementation of a compulsory procurement system in the countryside with a low and stable purchasing price. After 1979, despite the significant increase in grain purchasing price offered to the producers, the price for urban grain supply was stabilized involving an increasing government subsidy. As a result, for 32 years between 1952 and 1983, the retail price level for rationed grain increased only 22.6% in total.

The same retail price figure for food grain resold to the rural area is also available. It is the price offered by the official food marketing agency under the rationing system. The main differences compared with that in urban areas are, (1) grain resold to the rural area is only available to a small proportion of the rural population. It is targeted to those people engaged in specialized agricultural production other than grain growing, for example, those people who are specialized in growing cash crops, fishing and afforestation. (2) The retail price level for grain sold back to the countryside was not stabilized as much as in the urban areas. It was adjusted according to the changes in purchasing prices for agricultural products. The price level increased 42.8% during 1952–83, which is about twice the rise in prices for urban areas. (Appendix Table 1.12). Since the amount of grain sold back to the countryside was only 5–10% of the annual grain output for the last three decades (*TJNJ 1985*, p482) and targeted to a small group of non-grain producers, most of the rural population got food supply from the basic production units which was valued in official purchasing price. It can therefore be argued that the purchasing price is the main price signal determining the food grain consumption in rural areas. This can be further justified as follows. Given the very low income level most peasant families possessed for most of the time since the 1950s, saving a proportion of food grain to make some cash available is one of the few possible solutions for financing their other basic needs, such as clothing, health service, paying for their children's education and current production costs. Consequently, changes in purchasing price reflects the true or opportunity cost of food grain consumption. Although there seem to be strong arguments for using the purchasing price as the main price signal for rural food grain consumption, the selling price is also used in the empirical analysis as an alternative. In both the rural and urban sectors, an inverse relationship with the dependent variable is expected for the price variable.

Since differences in urban and rural sectors have been identified for several factors in consumption behaviour the estimation of demand for food grain is conducted separately for each group. It is therefore necessary to clarify the statistical definition of these two sectors. The distinction between the urban and rural sectors is based on the division of the administrative jurisdiction system practised in China. According to the SSB (see *TJNJ 1985*, p657), the urban population refers to the permanent residents in cities and towns, and the rural population therefore refers to the remainder in counties. The setting up of a city in China has to be recognized by the central government, and towns to be approved by the provincial governments. Before 1963, the norm for defining a town was the location of a township government which has a minimum total population of 2,000 in which more than 50% should be non-agricultural residents. Non-agricultural residents in China's countryside refer to those rural populations which are not engaged in agricultural production. Therefore their basic subsistence goods (including food grain) are rationed by the official marketing system rather than distributed by the basic production unit. This definition of a town was changed in 1964 to either a minimum number of permanent residents of 3,000 with more than 70% of non-agricultural population, or a minimum population of 2,500 in which non-agricultural residents account for more than 85%. Since 1984, either a township with a total population of less than 20,000 but where the township government located has more than 2,000 non-agricultural population, or a township with a total population of more than 20,000, in which more than 10% are non-agricultural residents, where the township government is located, can be categorized as a town. Therefore, the population in cities and towns of the nation has increased considerably from 1963 to 1984.

6.3.2 Food consumption - model estimation and results

Estimations for the consumption of food grain products have been carried out for urban and rural sectors and for the nation as a whole. Except for the figures of grain outputs and purchasing price, data used in the analysis is from *ZCWTZ* published by the SSB. The time-series data used in modelling are for 1952–83 and these are presented in Appendix One. Both linear and log-linear functional forms were tried, the results presented relate only to the linear function.

6.3.2.1 Urban consumption

The empirical model for urban food grain consumption, explained variations in the dependent variable, total food grains consumption in cities and towns (measured in 10,000 tonnes milled) in terms of:

- *-Total grain output in the previous year (100,000 tonnes unmilled)*
- *-Total population of cities and towns (100,000 persons)*
- *-Either per capita disposable income of citizens in cities and towns (yuan) or per capita expenditure on consumer goods of citizens in cities and towns (yuan)*
- *-Food grain retail price index in cities and towns (1952=100).*

Equations 6.1–6.3 in Table 6.3 show the results for aggregate urban consumption of food grains. All the coefficients have the expected signs. The signs of the estimated coefficients on grain output, population, income and consumer expenditure show that urban food grain consumption responds positively to changes in grain output, urban population growth, personal disposable income and cash expenditure on consumer goods. In addition, food grain consumption of the urban sector responds negatively to variations in the selling price of grain products. These results bear out the expected theoretical relationships. The magnitude of the estimated coefficients indicates that: for every additional 100,000 tonnes output of milled grain, urban demand increases between 2,100 and 2,900 tonnes; each extra 100,000 people stimulates additional grain demand of between 130,000 and 154,000 tonnes which is 130–153 kg per person on average; an improvement in average per capita disposable income of 1 yuan induces 10,500 tonnes more food grain demand (equation 6.1); and similarly, an increase of average per capita consumer goods expenditure by 1 yuan stimulates 13,200 tonnes of extra food grain consumption (equation 6.2); and a 1% increase in the retail price index could reduce the urban grain consumption between 88,800–118,100 tonnes. These magnitudes seem reasonable.

Statistical evidence in equations 6.1–6.3 shows that most of the estimated coefficients are highly significant statistically for they have passed the t-test at the 10% level of significance or better. The adjusted R^2 of 0.96 for all three equations indicates that the model specification, in spite of its limitations, successfully explains most of the historical variation in food grain consumption of the Chinese urban sector. This is further supported by the F-test which is significant at the 1%

Table 6.3 Urban food grain consumption

Equation	In aggregate term 6.1	6.2	6.3		In per capita terms: 6.4	6.5	6.6
Intercept	1,063.30** (1.95)	1,117.88** (2.28)	1,243.21** (2.42)	Intercept	2.71*** (5.89)	2.71*** (5.75)	2.99*** (9.65)
Grain output (100,000 tonnes)	0.21** (1.71)	0.23** (2.02)	0.29*** (3.25)	Grain output (0.1 tonnes head)	0.24*** (2.67)	0.23*** (2.70)	0.19*** (2.71)
Urban population (100,000)	1.30*** (4.64)	1.32*** (4.98)	1.54**** (9.74)	Grain price index	−0.01*** (−2.87)	−0.01*** (−2.66)	−0.01*** (−4.53)
Grain price index (1952=100)	−8.88* (−1.41)	−10.45** (−1.83)	−11.81** (−2.13)	Disposable income	−0.0003 (−0.81)		
Disposable income (yuan/head)	1.05 (1.00)			Consumer expenditure		−0.0004 (−0.79)	
Consumer expenditure (yuan/head)		1.32 (−1.01)					
Price elasticity	−0.38	−0.44	−0.50	Price elasticity	−0.58	−0.58	−0.58
Income elasticity	0.13			Income elasticity	−0.05		
$\bar{R}^2$	0.96	0.96	0.96	$\bar{R}^2$	0.37	0.37	0.38
F	200.57	200.71	267.15	F	7.09	7.07	10.43
D-W	0.95	0.99	1.02	D-W	0.79	0.78	0.73

Notes: Figures in the parentheses are the t-ratios of the estimated coefficients
***Significant at 1% level
** Significant at 5% level
* Significant at 10% level

level and implies that the model gives a good fit to the data. However, there are two major problems inherent with the aggregate urban demand estimation.

First, the coefficients on both specifications of the income variables are not statistically significantly different from zero. Both failed to pass the t-test at the 10% level of significance. There are two possible explanations for this. One is that because the urban food grain supply was under the monopoly control of the rationing system, in which the norms for food grain distribution were based on the numbers of the urban population, its distribution and the employment status (see section 6.2), logically, there was therefore no mechanism for a direct connection between urban food grain consumption and either personal disposable income or expenditure on consumer goods. The alternative explanation is that the high standard deviations of estimates of these two coefficients might be caused by collinearity with other independent variables. There is some evidence for this in the simple correlation matrix constructed, see Matrix 3.8, Appendix Two). To clarify this matter, urban food grain consumption was re-estimated in per capita terms with results shown by equations 6.4–6.6. A comparison between the aggregate equations and the per capita equations demonstrates that population growth is the most critical variable in determining growth in total urban food grain consumption. Evidence for this is substantial decline in values of R^2 (from 0.96 for the aggregate model to 0.37 in the per capita model). Despite removing the population variable there was no statistical improvement in the estimated coefficients on either of the income variables. Both once again failed to pass the t-test at the 10% level of significance. Thus the first explanation is favoured. That is, the food grain rationing system sterilized any impact of personal income growth on urban food grain consumption. This argument is further supported by the estimated elasticities derived from the equations. While price elasticities of between –0.38 – –0.58 have been obtained, the extremely low value of income elasticity (from –0.05 in equation 6.4 to 0.13 in equation 6.1) confirms that income growth was not an important factor in influencing the urban food grain consumption in the period up to 1983. As was pointed out in section 6.3 above, it could not be expected that the standard results of neo-classical demand theory would apply to the supply constrained circumstances of China. In such circumstances it is expected that income effects are likely to be understated; consumer pressure through higher incomes is not manifested as higher consumption during a time of shortage and may not result in higher prices either, due to the restrictions of the controlled price system. These problems are likely to distort estimated price effects to a smaller extent. There is no substantive trend in prices and there are few restrictions on price variation, particularly in the upward direction, influencing consumption.

The second analytical problem was that of serial correlation in the error terms of the equations. As can be seen from Table 6.3, the values of the D-W statistic for all the equations were close to unity, which indicates clearly the presence of positive auto-correlation in the unexplained residuals. This implies the ordinary least squares estimates of the regression coefficients are still unbiased but no longer have minimum variances. It is generally believed that the presence of serial correlation indicates a wrong specification no matter how high the R^2 may be. The usual responses to this problem are to change the model specification by adding new variables to the equations estimated; change the functional form; or change the estimation technique from ordinary least squares to Maximum Likelihood methods (Beach and Mackinnon 1978, pp51–58). Unfortunately, none of these removed the problem. The presented results were the best which could be achieved, it must therefore be borne in mind that the estimates have larger than desirable standard errors.

6.3.2.2 Rural consumption

The consumption relationships for the rural areas were of essentially the same form *mutatis mutandis* as for urban demand. Only in the definition of the price variable was there a significant difference. Two formulations were tried, an index of food grain retail prices in the countryside and a weighted average of prices of quota purchasing, over-quota purchasing and purchasing with negotiated prices. Both indices were based on 1952.

As in the investigation of urban consumption, the rural demand estimations were performed both in total and on a per capita basis. The results were presented in Table 6.4, in which equations 6.7–6.9 are for the aggregate consumption and equations 6.10–6.12 relate to per capita consumption. The results of the aggregate analysis were not very satisfactory. Coefficients on the population, food grain price index and consumer expenditure variables were not statistically significant based on the t-tests. The magnitude of the coefficients for population, and disposable income varied considerably between the alternative specifications in equations 6.7 to 6.9. These symptoms, together with the simple correlation matrix (see Appendix Matrix 2.9) suggest that the explanatory variables were highly inter-correlated. The presence of multi-collinearity makes it very difficult to assess the unique contribution of each of the independent variables and consequently the structural questions cannot be answered (see section 4.3.2). Therefore, the succeeding discussions of the empirical results concern the equations specified on a per capita basis, 6.10–6.12.

All the coefficients in these equations are statistically significant. The R^2 and F-tests are satisfactory and the coefficients less dependent on the model specification, all of which suggests that the multi-collinearity problem is largely removed by adopting the per capita basis. Rural grain consumption is positively influenced by grain availability and by disposable income (however measured) and negatively influenced by grain price specifications. The magnitudes of the estimated co-efficients indicate that a 100 kg increase in per capita grain output (unmilled) results in a 39 kg increase in per capita consumption (milled). A 1 yuan increase in per capita personal income induces between 2 and 5 kg more food grain consumption per head. The proportionate price responses derived from these models are between –0.21 and –0.33 and the corresponding income coefficient of demand is between 0.07 and 0.16. This price response is lower than for the urban areas – which was expected. The income response is higher, which was also expected. This latter result deserves further discussion.

Most of the rural food grains are produced and consumed as self-subsistence and is therefore not directly affected by the personal disposable cash income as used in the estimation (see section 6.3.1). However, it is expected that there will be a close relation between rural income and food grain consumption for the following reasons. First, it is recognized that due to the significant difference in the food grain distribution system between rural and urban sectors, the extent of rural grain consumption was lower than that in urban sector, both in quantity and quality terms (see section 6.3.1). It is also recognized that personal disposable income for the rural population was extremely low and there were few public services available in rural

Table 6.4 Rural food grain consumption

	In aggregate terms				In per capita terms:		
Equation	6.7	6.8	6.9		6.10	6.11	6.12
Intercept	293.07***	191.90***	260.01***	Intercept	0.75***	0.87***	0.82
	(3.53)	(3.47)	(3.45)		(6.66)	(5.53)	(6.96)
Grain output (100,000 tonnes)	0.48***	0.38***	0.44***	Grain output (0.1 tonnes head)	0.39***	0.51***	0.50***
	(9.16)	(7.95)	(9.58)		(9.71)	(11.43)	(13.25)
Rural population (100,000)	−0.22	−0.20	−0.09	Price index official grain purchased	−0.002***		
	(−0.68)	(0.87)	(−0.31)		(−5.01)		
Price index official grain purchased		−0.62**		Food grain retail price index		−0.005***	−0.004***
		(−2.40)				(−4.81)	(−5.20)
Food grain retail price index	−0.70		−0.41	Disposable income (yuan/head)	0.005***		0.002***
	(−0.45)		(−0.31)		−6.20		(5.63)
Disposable income (yuan/head)		2.32***	0.99**	Consumer expenditure (yuan/head)		−0.002***	
		(3.42)	(2.33)			(4.14)	
Consumer expenditure (yuan/head)	0.44						
	(0.95)						
Price elasticity	−0.07	−0.10	−0.04	Price elasticity	−0.21	−0.33	−0.27
Income elasticity		0.12	0.05	Income elasticity	0.16		0.07
R^2	0.98	0.98	0.99	R^2	0.95	0.93	0.95
F	485.68	688.02	568.20	F	180.83	139.63	187.70
D–W	1.54	1.68	1.51	D–W	1.62	1.35	1.45

Notes: Figures in the parentheses are the t-ratios of the estimated coefficients
*** Significant at 1% level
** Significant at 5% level
* Significant at 10% level

areas. So, quite apart from facing the problem of maintaining their farming activities, peasant households have total responsibility for all their basic needs. This includes food, clothing, housing, social service payments (typically health service), fees for children's education and expenditure on social relations (social life is still largely determined by traditional customs). As a considerable proportion of the basic needs have to be covered by cash income, peasant households must carefully balance all consumption items in order to satisfy their minimum level of basic needs. In doing so, one possible way of creating a cash supply was to squeeze food consumption to enable grain selling. As a result, like the price response behaviour discussed in section 6.3.1, while higher disposable income enabled peasant households to retain more grain for their own consumption, a significant proportion of grain was saved and sold when less cash income was obtained. In these circumstances it is not so surprising that the coefficient on disposable income is significant but small, it is summarizing effects which have both a positive and negative effect on grain consumption.

6.3.2.3 Total food grain consumption

The aggregate consumption of food grain combines both the urban and rural sectors. This analysis provides a general picture of food grain consumption of the nation. The results are summarized in Table 6.5.

Table 6.5 Total food grain consumption

Equation	6.13	
Intercept	93.01	(0.70)
Grain output (100,000 tonnes)	0.24***	(3.47)
Population (millions)	1.22***	(2.97)
National per capita disposable income (yuan)	3.58***	(5.03)
Retail price food products (yuan)	−2.56**	(−1.84)
$\bar{R}^2$	0.99	
F-ratio	1,039.3	
D-W	1.83	

Notes: Figures in the parentheses are the t-ratios of the estimated coefficients
***Significant at the 1% level
**Significant at the 5% level

The basic features of the estimated result in equation 6.13 confirms that, under the Chinese circumstances, consumption of food grain is fundamentally determined by the production level achieved, the growth in population, the improvement in per capita disposable income and the selling price of food grain products. Moreover, during the last three decades, food grain requirements responded positively to domestic production, population growth and disposable income improvement, and negatively to the selling price of the product. The robustness of the result is borne out by the conventional statistical tests, in which the high significance of the estimates of the independent variables are demonstrated by the t-test which passed either at 1% or 5% significance level. The high value of R^2 shows the goodness of fit condition of the equation and is further supported by the F-test, which passed 1% significance level successfully. In addition, the value of the D-W statistic implies that there is no evidence of serial correlation in the unknown residuals of the equation.

The results indicate, as expected, that the consumption of food grains in China is very unresponsive to price and income changes. The proportionate price response was –0.38 to –0.58 for the urban sector (Table 6.3), –0.21 to –0.33 for the rural sector (equations 6.10–6.12, Table 6.4), and –0.21 for aggregate consumption (equation 6.13). The conventional explanation for these low responses is based on the idea that food grains are a basic need for consumers. In China's case, another reason is that, since the market for grain and other major food products were under monopoly control by the government in which prices were fixed and the market link between production and consumption was severed, consumers' freedom of choice was very restrained. The proportionate income response was found to be not significantly different from zero for the urban sector, in the range of 0.07–0.16 for the rural sector (equations 6.10, 6.12, Table 6.4), and 0.17 for the national aggregate demand (equation 6.13). These are consistent with results from other empirical research. For example, an income elasticity of demand for total grain of 0.25 was used by World Bank in projecting the growth in per capita grain requirements for the period 1980–2000 (World Bank 1985b, p19); the income elasticities of total grain consumption for income groups in urban China was estimated at the range of 0.05–0.35 by Kueh (1984, p1264). The results are also consistent with those for other comparable countries which have consumption constrained by grain availability. For example, the correspondent elasticities were 0.35 for Taiwan, 0.25 for India and 0.14 for Singapore (Kueh 1984, p1264).

Since these results contradict the expectation of "a high income elasticity of demand for grain at the prevailing low levels of income per head" expressed by Walker (1984), it is worth searching for an explanation. There are two factors which may explain this disparity. One is that the income improvement itself does not automatically drive up the demand for food grain due to the imposed rationing system. This was clearly shown by the analysis of the urban sector. Thus the response coefficients estimated are not the unconstrained income elasticities of neo-classical economics referred to by Walker; rather they measure the responsiveness of consumers to income growth in highly constrained circumstances. The other factor is that growth in income raises the indirect demand for food rather than grain directly. This is shown by Tang's projection of grain demand (medium) for 1977–2000, in which 0.25 was taken as the income elasticity of direct demand, 2.0 for indirect demand, and 0.60 for aggregate demand (Tang and Stone 1980, p36).

In summary, the empirical results from the food grain demand analyses have shown that, under China's circumstances, consumption of food grains is fundamentally determined by the grain production, population growth, changes in personal disposable income (not in urban sector for the reasons exposed in section 6.3.2.1), and the prices of grain products.

6.3.3 Grain for feed

6.3.3.1 Basic features of grain consumption

There are three main characteristics inherent in China's livestock raising and feed grain demand. First, the pig is the predominant domestic animal. In 1979, 320 million head of pig were raised (year-end figure) (*TJNJ 1985*, p267) which made up 42% of the total number in the world (Bureau of Planning 1981, p78). In 1982, pork accounted for 94% of the total meat output (in Chinese statistics, meat output covers pork, beef and mutton, poultry meat is excluded) and beef and mutton only 6%. For the same year, the world average is pork 39%, beef and mutton 38%, poultry 21%, and others 3%. (He 1985, p558). The retail selling value of pork in 1983 was 17.1 billion yuan, which is 45, 29.5, 14.3 and 6.2 times that for beef, mutton, poultry and fresh eggs respectively (*ZCWTZ 1983*, pp91–92).

Second, the livestock industry is concentrated in the south and east of China where the major farming areas which are sources of feed grains may be relied upon. Currently 90% of the meat is produced in arable farming regions and only 10% from the vast grazing area in the north and west regions (He 1985, p2). Consequently, animal raising, especially pig and poultry raising, is mainly part of the peasant household's sideline occupations. Feed grain is not distinguished from food grain. It is mixed up with other grain sources in the households and consumption between humans and animals are thus indistinguishable. The published annual statistics for feed grain are estimated figures derived by multiplying the feeding standards for different animals with the yearly average number of the corresponding livestock categories (SSB 1986, p60).

Third, the proportion of total grain used for feed is low and varies considerably. In 1952, feed grain made up only 8% of the total grain consumption of the nation and that for 1982 was 15%. (The Coordinating Office, the National Agricultural Zoning Committee 1985, p52 (hereafter NAZC 1985)), while for the world on average during 1975–79, grain for feed made up 43% of the total, 0.6% higher than food grain directly consumed by human beings (Zhang 1985, p493). From 1952 to 1983, feed grain in China increased from 9.8 to 44.2 million tonnes, a net growth of 3.5-fold (Appendix Table 1.10). Within this growth, however, the use of grain for feed is highly sensitive to changes in per capita grain availability. When grain availability is good, a higher proportion of grain would be allocated for indirect consumption. But when it is poor, the indirect consumption is squeezed to safeguard direct food consumption. Another important factor in influencing the feed grain requirement is the profitability of the livestock industry. Peasant farmers are generally very sensitive towards the changes in price ratio between grain and animal products, especially between grain and pork. When the price ratio is in favour of pig raising, peasant households usually raise more pigs and curtail the direct selling of grain products. The situation is reversed when pig raising becomes unattractive (He 1985, p569). For example, in 1984 after several bumper harvests achieved in corn production in Jilin Province of northeast China, an unprecedented volume of official purchasing took place and an overstorage condition was created for corn products in that province. However, since the price ratio of meat and grain dampened peasant farmers' enthusiasm for pig raising, instead of using more corn to raise more pigs, the number of pigs was actually reduced and farmers sold most of the surplus corn products directly to the official marketing system. This created trouble for the provincial and central governments by the end of the year. Other indicators of the volatility of feed grain consumption are provided by the national aggregate statistics, which show that a 40% feed grain reduction was registered in three years between 1958–61 and a 38% increase occurred during 1977–82 (Appendix Table 1.10).

6.3.3.2 Model specification and results

Theoretically, the demand for feed grain is determined by the supply and demand for livestock products as well as the price ratio between animal products and feed grain. Practically, several issues require clarification before the empirical function is specified. First, given that beef and mutton are mainly produced in the northern grasslands and feed grain is not a significant feed component, it is not necessary to introduce the consumption of beef and mutton as an explanatory variable. Second, although poultry production is the second most important category of animal raising after pigs for peasant households in farming areas, and poultry meat has been the second most important meat consumption item for most Chinese people, since there is no time-series statistics on either poultry meat production or consumption, it is impossible to measure its relation to the feed grain demand. Third, unlike that of Western patterns of food consumption, dairy products, such as milk, cheese and butter, have not been significant items for most Chinese people's diet (although it is the case for some minority groups from, for example, Mongolia, Tibet and Uygul). In 1983, total milk production in China was around 1.8 million tonnes with per capita consumption of only 1.7 kg, which is very low compared with 10 kg in Malaysia and 103 kg in the United States (World Bank 1985a, p13). The demand for feed grain in dairy production is therefore insignificant. Bearing all these in mind, feed grain demand in this study is empirically specified as a function of the total consumption of pork and eggs, the price ratio of livestock (or pork) and feed grain products and a time trend variable to cover all other feed grain consuming items except from that of pig and layer raising industries. Logically, positive relationships between feed grain consumption and the consumption of pork and eggs and the time trend variable are expected. In addition, a positive relation between feed grain consumption and the price ratio between animal products and grain products is also anticipated. This formulation is thus based on the assumption of rational behaviour in animal raising. The estimated results are presented in Table 6.6.

Like the other empirical results presented so far, equations 6.14–6.17 of Table 6.6 were estimated using the method of OLS. As expected, results from these equations show that the consumption of feed grains is positively related to all the included independent variables, i.e. total consumption of pork, eggs, other grain consuming animals, and the price ratios between livestock and grain products and

Table 6.6 Feed grain consumption

Equation	6.14	6.15	6.16	6.17	6.18	6.19	6.20	6.21
Intercept	154.22*	53.06	−380.48	−177.12	1,149.94*	190.53	201.36	219.02
	(1.35)	(0.06)	(-0.78)	(0.36)	(1.66)	(1.02)	(0.45)	(0.47)
Pork consumption (10,000 tonnes carcase wt)	3.83***	2.38***	1.55**	1.58**	1.96**	1.72***	1.52***	1.52***
	(20.39)	(8.95)	(2.36)	(2.37)	(4.75)	(4.52)	(2.61)	(2.61)
Egg consumption (10,000 tonnes)			4.75*	4.55*			1.17	1.16
			(1.45)	(1.38)			(0.46)	(0.45)
Price ratio of animal products to grain			381.87				−7.75	
			(0.89)				(0.20)	
Price ratio pork to grain				194.47				−22.73
				(0.48)				(0.60)
Time		54.21***	50.55***	49.34***		67.97***	65.37***	65.30
		(6.20)	(5.29)	(5.17)		(4.70)	(4.12)	(4.12)
R^2	0.93	0.97	0.97	0.97	0.19	0.82	0.81	0.81
F	415.57	486.52	248.82	243.54	8.35	69.96	31.17	31.10
D-W	0.52	0.69	0.81	0.77	1.80	1.86	1.83	1.83

Notes: 1. Equations 6.14–6.17 were estimated with OLS and equations 6.18–6.21 with Maximum Likelihood (ML) methods, both are in linear function form. 2. Figures in the parentheses are the t-ratios of the estimated coefficients.
***Significant at 1% level
**Significant at 5% level
*Significant at 10% level

between pork and grains. Statistically, coefficients for the pork, eggs and time trend variables are significantly different from zero. The R^2 statistic shows that more than 90% of the variation of demand for feed grain has been explained by the model, this is supported by F-test for all of them have passed 1% significance level successfully.

The magnitude of the coefficient on pork consumption in equation 6.14 indicates that 3.83 kg feed grain is used for the conversion of 1 kg pork. The corresponding estimate from equation 6.15 is 2.38 kg. Both of these are judged acceptable. Although the conversion rate for pigs in China was often assumed to be 4–6:1 (in terms of unmilled grain, see World Bank 1985b, p23), a conversion rate of around 3:1 is appropriate after taking into account the reality of other feeds used by peasants in pig raising. Apart from feed grains, animal feeds also include by-products of grain milling and high protein meal or cake from oilseed. The magnitude of the correspondent estimates in equations 6.16 and 6.17 are significantly below 2 and this is not considered acceptable. The conversion rates estimated for egg production, around 4–5:1, are assumed to be reasonable in China's circumstances, in view of the small size and non-commercial orientation of peasant household's production condition. Estimates of the coefficients on the price ratios are not significantly different from zero statistically, for which the problem of data inaccuracy and multi-collinearity are suspected as two possible reasons.

The most serious issue inherent in the results in equations 6.14–6.17 is the strong evidence of positive correlation between error terms. Bearing in mind that if the disturbances of a linear regression model are correlated, the coefficient estimates with ordinary least squares are inefficient although still unbiased. The Maximum Likelihood (ML) method was also tried with results presented in equations 6.18–6.21 of the same table. (The method used is due to Beath and MacKinnon 1978, pp51–58.) As can be seen from the results in equations 6.18–6.21, while the problem of serial correlation was successfully overcome, the magnitudes of coefficients for pork and egg consumption are unacceptably low. In addition, the negative signs for the two price ratios are also unreasonable. Consequently, equation 6.15 is assumed to be the best for further application.

6.3.4 Grain for seed

6.3.4.1 Basic features of the dissemination of high-yielding varieties (HYVs)

As elsewhere, much of China's science and technology effort in agriculture has been concentrated on developing high-yielding plant varieties, especially in cereals since the second half of the 1950s. The desirable characteristics of HYVs pursued in China are high yield, disease resistance and early maturity (which enables the multiple growing index to increase). This is the critical component of the "three-in-one combination of HYV seed, water and fertilizer" which described the basic features of the "Green Revolution" in Chinese agriculture since the early 1960s (Hussain 1986, pp25, 30).

To co-ordinate the national seed policy and planning, the National Seed Corporation (NSC) was set up under the MAAF in 1978. The prime missions of the

NSC is to regulate international trade in seeds, co-ordinate domestic seed production and distribution and to provide technical guidance for provincial and county seed companies. There are 2,300 local seed companies which are responsible for the multiplication, processing, storage, certification and distribution of seeds to be used in production. Multiplication of stock and certified seed is either done by special seed farms operated by local seed companies, or by contractual arrangements with basic production units and peasant households under which the seed is returned to the company for processing and distribution (World Bank 1985b, p40). Nevertheless, for a huge country with vast and varied farming systems, the diffusion of a HYV is critically dependent upon its adaptation to the local conditions as well as its slot in the seasonal cycle. Moreover, huge supervision and technical services are needed in addition to the requirements for storage, processing and transportation facilities. This is why until 1983–84, the supply of seed produced under the NSC supervision totalled only 2.5 million tonnes and made up less than 13% of the total seed requirement. Most peasant farmers have therefore still to rely on retained seed for reproduction (World Bank, 1985b, p40; Hussain 1986, p26).

Great achievement has been made from the continuing efforts in seed variety improvements. The creation and dissemination of high-yielding semi-dwarf rice varieties has improved the yield from 3 to 3.7 tonnes/ha in the middle of the 1950s to 4.5 to 5.2 tonnes/ha in the early 1970s. During 1976–83, the aggregate planting area of 34 million hectares of hybrid rice varieties contributed to more than 25 million tonnes of additional output which thus registered 0.75 tonnes/ha further yield improvement (Huang 1985, p503). For wheat cultivation, the introduction of a semi-dwarf variety in 1962 has successfully curtailed the damage created by scab which brought a net output loss of 6 million tonnes in the early days (Niu 1985, p344). In corn production, the "Zhongdan No. 2" variety which is able to resist three common diseases was cultivated on 1.74 million hectares in 1983. This variety has a yield improvement of up to 20% over conventional varieties (Lu 1984, p3). For sweet potato, which is another staple food crop in China, the "Xushu No. 18" HYV can improve yield by 3 tonnes/ha (calculated in grain equivalent), and it had been disseminated to a growing area of 1.3 million hectares by 1983 (Huang 1985, p503).

6.3.4.2 Model specification and results

The technical complementarity between seed and land inputs is well established, thus growing area is conventionally regarded as the fundamental element in determining the seed requirement in grain production. Another factor usually considered is a time trend variable representing technological improvement. The general expectation is that a continuous decline in seed requirement per unit area can be achieved from efforts in breeding, testing and sowing seeds (Colman 1972, p79). Thus, the demand of seed is usually specified as a function of grain growing area and a time trend variable.

Since the figures for seed grain use in China are derived by summing the product of annual growing areas of individual cereal crops with correspondent per unit seeding standards (SSB 1986, p60), the positive relationship between seed use and growing area is guaranteed. Despite the tremendous efforts which have been made in technological improvement, the use of seed grain has increased from 7.4 million tonnes in 1952 to 23.5 million tonnes in 1983, whilst the total growing area decreased by 10 million hectares (Appendix Table 1.2 and 1.10). The requirements per hectare have therefore risen from 0.06 tonnes to 0.22 tonnes and the proportion of seed demand in total grain output has risen from 4.5% to 6.1%. This strong upward movement is mainly the result of two factors. The first is the effect of crop variety substitution which has enabled denser planting. For example, the introduction of semi-dwarf rice and wheat varieties since the 1960s and hybrid varieties for rice, corn and sorghum since the early 1970s have increased the cropping density to 30–50% on average, which requires an additional 30% of extra seed. The second comes from the changes in crop composition. While a marginal reduction in total grain growing area took place, the planting area for rice, wheat and corn, which require higher seeding rates, has been increased from 66 to 81 million hectares between 1952 and 1983. Correspondingly, the growing area for crops with lower seeding requirements, like soybean, sorghum and millet, decreased (Appendix Table 1.2). Since the effect caused by these two factors is far greater than those others contributing to a net reduction in seed requirements, a positive relationship is expected between seed demand and the time trend variable in this analysis. The estimated result is presented in Table 6.7.

As shown by Equation 6.22, the coefficients on growing area and the time trend variables have the expected positive signs and both are significant. The implied rate of seed application of 0.189 tonnes/ha is well in line with quoted technical coefficients (0.06–0.22 tonnes/ha during 1952–83), and the coefficient on the time trend variable which indicates a 0.66 million tonne annual increase in seed requirement is also acceptable. However, despite a high value for the adjusted coefficient of multiple determination ($R^2 = 0.95$) and the F-test which is significant

Table 6.7 The demand for grain for seed

Equation	6.22	
Intercept	–1,734.5**	(–1.99)
Grain growing area	+1.89***	(2.79)
Time trend	66.42***	(17.62)
$\bar{R}^2$	0.95	
F-ratio	251.6	
D-W	0.57	

Notes: Figures in the parentheses are the t-ratios of the estimated coefficients
***Significant at the 1% level
** Significant at the 5% level

at the 1% level, the low D-W value of 0.57 suggests a positive correlation between error terms. This was not pursued further.

6.3.5 Industrial consumption of grain

According to the statistical definition published by the SSB (1986, pp28, 60), industrial consumption of grain covers all the industrial products which utilize grain including cakes and biscuit production in the food processing industry, wine, beer and spirits, ethyl alcohol, soysauce and the vinegar making industry, and other outlets such as pharmacy and sizing in the textile industry. From the point of view of economic analysis, the industrial use of grain can be identified as two separate categories. One is the grain requirement for making wine, biscuits and cake. The majority of industrial demand is incurred in this category. Since drinks and processed foods are not part of the basic needs for most people's daily life given the average standards of life, any increase in demand is mainly determined by the improvement in per capita disposable income. The second category of industrial grain includes all other industrial uses. The products included in this group are relatively small and are directly or indirectly linked to the basic needs of people's daily life. They are income inelastic and the increase in demand is fundamentally driven by the grown in population. The specification of industrial grain demand in this book takes both categories into account. Consumption of individual items are taken as the indicators for the first category, and a time trend is considered as the proxy for the second. A positive relationship between grain demand and all the selected independent variables is expected. The estimated result is presented in Table 6.8.

Table 6.8 The industrial use of grain

Equation	6.23	
Intercept	125.67**	(2.20)
Consumption of biscuit and cake	0.52	(0.41)
Consumption of wine	1.64***	(7.71)
Time	6.41	(0.75)
$\bar{R}^2$	0.95	
F-ratio	136.8	
D-W	1.51	

Notes: Figures in the parentheses are the t-ratios of the estimated coefficients
***Significant at the 1% level
** Significant at the 5% level

The positive signs for all the independent variables in equation 6.23 confirms the general expectations discussed above. Statistically, the high value of R^2 shows that about 95% of the variation in industrial grain demand has been explained by the model. This is supported by the high and significant F-statistic. However, half of the estimated coefficients failed to pass the t-test at the 10% significance level which makes them less reliable. (The D-W value of well below 2 is inconclusive regarding the presence of serial correlation in the error term.)

Given than a 0.4 kg grain coupon is required in China to buy 1.0 kg of cake and 0.6 kg is required for 1.0 kg of biscuits, the coefficient on the consumption of cake and biscuits which indicates a requirement of 0.52 tonne of extra grain per tonne of biscuits and cake in equation 6.23 is regarded as appropriate. The technical co-efficient for wine making varies drastically between beer and spirits. For beer making, about 10.0 kg of beer can be produced from 1.0 kg of grain input. Meanwhile, 1.0 kg of spirits usually requires more than 1.0 kg of grain. Bearing in mind that the majority of wine produced in China is not spirits, 1.64 tonnes of grain for making one tonne of wine derived from equation 6.23 is regarded as unaccept-able. These rather unsatisfactory results for biscuits, cake and wine suggest the model should not be used for policy analysis work, although the equation might predict industrial grain consumption quite satisfactorily.

6.4 Summary and Conclusions

In this chapter, the basic features of China's grain distribution system was outlined. This was followed by the specification and estimation of the consumption relations for food, feed, seed and industrial usage of grain.

The demand for food grain is the most important part of grain consumption because of its high share of the total. Results of both urban and rural demand consumption analyses show that, despite fundamental differences in the distribu-tional systems between urban and rural areas, in both cases quantities consumed are explained most by the domestic production level achieved, the growth in population and prices of grain (the selling price in the urban sector and the purchasing price in the countryside). The positive signs of regression coefficients of output and population growth demonstrate that food grain consumption responds positively to the changes in domestic production and population growth. Food grain consump-tion responds conventionally and thus, negatively, to the changes in price of grain. For consumers in cities and towns, there are a number of internal substitution possibilities within grains and between grain and other foods. For consumers in rural areas who are also grain producers the choice faced is between consuming and saving. The results demonstrate that both groups are economically rational in responding to price signals.

The results of the urban consumption analysis (both at the aggregate level and on a per capita basis) demonstrated that there was no strong evidence supporting a significant correlation between urban consumption of food grains and either personal disposable income or expenditure on consumer goods. It is believed that since the urban food grain supply was controlled by the rationing system, in which the amounts distributed were based on the population characteristics and the employment status, there was therefore no mechanism for a direct connection between income growth and food grain consumption. However, a significant correlation between the two was found in corresponding rural demand analysis. Given the extremely low level of disposable income possessed, squeezing food consumption to enable grain to be sold was one of the few ways for most rural families to meet expenditures on various other vital consumption items to achieve a minimum level of basic needs. Consequently, while higher disposable income level can encourage peasant households to retain more grain for family consumption, a significant proportion of grain was saved and sold when less cash income was obtained.

Putting the urban and rural demands for food grains together, it was shown in an analysis of total food grain consumption (equation 6.13) that, 0.24 million tonnes of increase in consumption of milled grain is stimulated by a one million tonnes improvement in output of unmilled grain. Also one extra person adds 122 kg additional food grains to total demand. Together with the evidence of the low price and income elasticities of demand, these magnitudes indicate that food grain consumption is not only dependent upon the consumer's income level and price signals but also, and indeed more importantly, on the production level achieved and the growth in population.

Turning to the use of grain for livestock feed the evidence shows that demand is directly determined by the consumption requirement of animal products and the changes in relative price of livestock products and feed grains. As demonstrated by the results in Table 6.5, feed demand responds positively to changes in consumption level of pork, eggs and a time trend variable which is a proxy for other grain consuming livestock products. The higher the consumption requirement for these animal products, the higher the demand for feed grain. In addition, the positive signs of price ratios of animal products and grain and between pork and grain indicate the economic rationality of peasant household's supply response behaviour in livestock production. Most peasant farmers are both grain and livestock producers, the choice for them is between selling grain products to the market directly or retaining them as input for livestock production.

The use of grain for seed follows the conventional pattern. Changes in seed requirements are determined by changes in growing area of grain crops and technological improvement. Because of the revolution in high-yielding varieties which resulted in a considerable change in crop composition (especially a shift from crops with lower seed requirements per unit to crops with higher seed rates) and in cultivation patterns (in particular the advent of closer planting), the effect of technological improvement in the nation has significantly increased seed requirements at the aggregate level.

While the industrial use of grain covers a large number of industrial products, two separate categories were identified in specifying an industrial demand function. One is the grain requirement for producing wine, biscuits and cake. This is the major part of industrial demand. The increase in consumption of these products is mainly induced by the improvement in personal incomes. The other category is the residual industrial demand. The attempt to estimate economic relationships for this rather heterogeneous group did not meet with success.

This completes the analysis of domestic grain production and consumption. It remains to put these two together and to review the rôle of foreign trade in maintaining a balance in China's grain economy.

Chapter Seven

Grain Imports and Exports

7.1 Introduction

China is not only the world's largest grain producer, but is also a major grain importer and exporter. Grain trade is a significant component in balancing domestic supply and demand. This chapter first outlines the basic trade pattern for China since the 1950s. It then narrows down to the specific features of grain imports and exports revealing the complexity of the nation's grain trade in the last three decades, and exploring the arguments concerning the changing motivation for trade. This is followed by an identification of the main themes for the periods of 1950s, 1961–65, 1966–77 and 1978–83.

Following the broad, qualitative discussion is an attempt to quantify trade functions. Models are estimated for rice exports, wheat imports and net trade of grain products. The hypothesis tested is that together with the price signals of the world market, both domestic supply and consumption are key factors in determining grain trade. Putting the trade functions together with the estimated demand functions completes a recursive econometric model system for the Chinese grain economy which can then be used to analyse possible future trajectories.

7.2 Review of the Trade Patterns

7.2.1 The institutional setting

Since the founding of the PRC, China's trade sector has been controlled by the central government. The Ministry of Foreign Trade was set up under the State Council and is responsible for decision-making concerning what, how, how much and where to import and export. These decisions were then executed by the various trade companies under the administrative system. Examples of such companies are the China National Grain and Edible-Oil Products Import and Export Corporation and the China National Native Produce and Animal By-Products Import and Export Corporation. The official exchange rate for visible trade was fixed by the Ministry of Finance, at two yuan (unit of Renminbi, the Chinese domestic currency) for one US dollar for a long period of time and was then adjusted to 2.8:1 at the end of the 1970s. Annual accounts both in terms of foreign exchange and domestic currency were drawn by the trade administrative system and companies following the general principle of trying to balance the value of imports and exports. Since most agricultural products were undervalued in domestic markets, the trade surplus (in terms of domestic currency) made by companies had to be submitted as revenue to the state treasury. At the beginning of the 1980s, the revenue from exporting agricultural products amounted to more than 200 million yuan per annum. For products, especially manufacturing goods which were overvalued, the deficit over the planned target (also in domestic currency) had to be declared at the end of the financial year and then filled by the state exchequer. Because of the isolation of the domestic sector and foreign trade, there was no direct economic linkage either between producers and exports or domestic dealers and imports. For the stabilization of trade, short-term to mid-term international agreements were usually concluded in the form of bilateral treaties. Typical examples are the agreements for agricultural exports with the Soviet Union made during the 1950s and wheat import treaties agreed with the United States, Australia and Canada in the late 1970s.

7.2.2 The basic pattern of total Chinese trade

The major characteristics of China's trade sector can be summarized as relatively closed, oriented to technological imports and erratic from period to period.

Closed trade stance Although her total trade value increased from 1.9 billion dollars in 1952 to 40.7 billion dollars in 1983, with an average growth rate of 10.3% per annum (*ZCWTZ 1952–1983,* p484), China has a relatively closed economy in which foreign trade has not been a major component in its GNP since it was isolated from the outside world for most of the three decades following the creation of the PRC. During the period 1952–83, the share of China's import and export value in the world market fluctuated between 0.6% (1977) and 1.8% (1959) (*ZCWTZ 1959–1983,* p489), which accounted for 4.5% to 5.9% in its GNP respectively (*TJNJ 1985,* p20 and *ZCWTZ 1952–1983,* p484). This is partly due to the large continental nature of the Chinese economy, but is also an outcome of conservative foreign trade policies (Surls 1982, p184). Nevertheless, because of the nature of China's economy in the early 1950s and the government's continuing drive for industrialization and modernization, foreign trade has always been seen as an important factor to foster economic development.

The motivation for trade Since the planned economy has been regarded as the basic nature of a socialist economic system, foreign trade is often characterized as the domain of a state monopoly in which imports are largely limited to filling shortfalls in domestic economic plans. In pursuing the aim of industrialization, the basic trade pattern for the last three decades is to extract products from the traditional sector (agriculture) for export to enable imports of manufacturing facilities and technical know-how from the developed world. Only in this way can the modern sectors be strengthened, especially heavy industry. As shown in Table 7.1, the export share of agricultural and sideline products and their processed goods made up 77% of the total export in the 1950s, 72% in the 1960s, 66% during 1970–77, and 47% for 1978–83. On the other hand, the import of producer goods made up more

Table 7.1 The rôle of agriculture in China's foreign exports (in 0.1 billion US dollars)

Year	Total export (a)	In which				
		Agriculture and sideline products (b)	% of the total (c)	Processed agricultural and sideline products (d)	% of the total (e)	(c)+(e)
1952	8.23	4.88	59	1.88	23	82
1957	15.97	6.40	40	5.03	32	72
1962	14.90	2.89	19	6.84	46	65
1965	22.28	7.37	33	8.02	36	69
1970	22.60	8.29	37	8.52	38	75
1975	72.64	21.50	30	22.61	31	61
1978	97.45	26.91	28	34.14	35	63
1983	221.97	35.29	16	59.20	27	43
1950-59	131.96	58.86	44.6	42.19	32.0	76.6
1960-69	194.08	62.00	31.9	77.38	39.9	71.8
1970-77	428.15	136.63	31.9	146.31	34.2	66.1
1978-83	1,065.83	197.24	18.5	300.40	28.2	46.7

Sources: Columns (a), (b) and (d) are from *ZCWTZ 1952-1983*, p490
Columns (c) and (e) are from *ZCWTZ 1952-1983*, p491

than 90% of the total during 1953–59 and more than 70% for the whole period except from 1960–65 (*ZCWTZ 1952–1983*, p507). This clearly illustrates the unimportance of importing consumer goods. Thus the rôle of agriculture in the development of China's national economy has not only been to feed the large population and to provide resources to build its industrial sectors, but it has also been a most important part of Chinese external trade.

Variability of trade As shown in Chart 7.1, China's foreign trade sector grew substantially in the 1950s, declined in the early 1960s, increased rapidly in the early 1970s, suffered a temporary set-back during 1974–76, and then has shown continuing rapid growth since 1977. The values quoted and plotted are nominal and thus the real growth in trade is considerably less than that depicted. The periodic fluctuations in total trade reflect the shifts in the leadership's attitude between an outward and inward looking economic policy. When economic development and modernization was stressed, effort was made to expand foreign imports of capital goods, new technology and industrial raw materials. However, when an inward looking policy was pursued, self reliance was emphasized and trade was therefore restrained. Furthermore, the shifts between inward and outward looking strategies were deeply

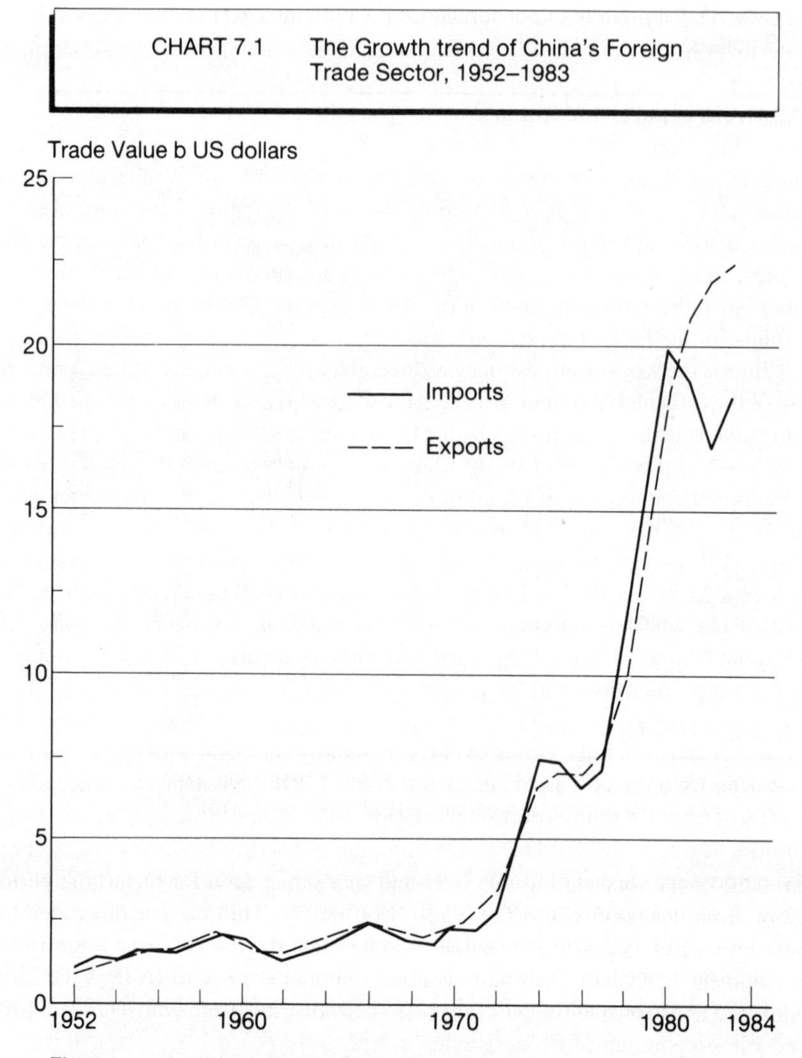

Figures used are from the foreign trade department, these are different from those of the Customs department. Source: ZCWTZ 1952–1983 p 484

rooted in the developments in the internal and external political climate. The expansion in the 1950s reflected the ambitions of the top leaders and a good relationship with the USSR as well as other countries in the Eastern Bloc. The reduction in the 1960s was caused by the domestic political chaos of the Great Leap Forward and Cultural Revolution and the breakdown in the friendship with the Soviet Union. Equally, the improvement of China's relationships with the United States and other Western industrialized countries and her re-emergence as a member of the United Nations made the expansion of trade in the early 1970s possible. The temporary set-back during 1974–76 was mainly caused by the domestic political struggle with a virulent campaign against Mr Deng Xiaoping.

7.2.3 Grain exports and imports

Coincident with the importance of food production in China's agriculture, food imports and exports are the major components of the nation's agricultural trade. Although there are many products involved in food exports, for example live animals, meat, eggs, fruits and vegetables, grain products, especially rice, are consistently the most important agricultural exports. Grain, mainly wheat, also accounts for the bulk of the nation's food imports.

China is the world's largest rice producer. Her total output accounted for nearly 40% of the total global output in 1982. Her share of world rice exports at 10.3% in 1980 ranked third, exceeded only by the United States (24%) and Thailand (21.6%). China's wheat production made up 12.4% of the total output of the world in 1982. Wheat imports amounted to 12.3 million tonnes, which is 12.4% of the total world imports in 1980 (Kueh 1984, p1247). The bulk of the nation's rice exports are destined for the rice consuming countries of southeast Asia, including Vietnam, Indonesia, Malaysia, the Philippines, Singapore and Thailand (Wong 1979, p451). Most of the wheat is imported from the United States, Canada, Australia and Argentina (Barnett 1981, p355). Further details of the historical developments of grain exports and imports in the last three decades are outlined below.

Grain exports China's grain exports are mainly characterized by the exporting of rice and soybean. As can be seen from Table 7.2, these two products accounted for 80% of China's grain exports in the 1950s, 74% in the 1960s, 72% for 1970–77, and 66% during 1978–83. Due to the continuing discrimination against soybean production since the end of the 1950s (for details see Chapter Four), the 9.7 million tonnes of soybean output in 1984 is still less than 10.2 million tonnes achieved in 1956. This static production caused the soybean exports to decrease over time from 10.9 million tonnes in the 1950s to 1.1 million tonnes during 1978–83. The proportion of soybeans in total grain exports decreased from 49% to 12% for the same period.

The reduction in soybean exports was complemented by the increase in rice exports from 0.3 million tonnes in 1952 to the peak of 2.6 million tonnes in 1973 (*ZCWTZ 1952–1983*, p496). Consequently, the proportion of rice in total exports rose from 32% in the 1950s to 57% in 1970–77 and 54% for 1978–83.

Grain imports Unlike other major grain importers like Japan and the Soviet Union, the basic purpose of China's grain imports is for the direct consumption of human beings, especially for feeding the urban population of the three metropolitan areas of Beijing, Tianjing and Shanghai. As China started to import large quantities of food grain in the early 1960s, wheat has always been a major component. Wheat imports in the 1960s made up 88% of the total grain imports, and it has been maintained at more than 80% since the 1970s (Table 7.2).

China changed its position from being a net grain exporter to net importer at the beginning of the 1960s. The whole period of the grain imports and exports since the 1950s can be divided into three different phases. The first phase is the 1950s. During this period, China exported 22.3 million tonnes of its grain products with a total net export of 21.5 million tonnes (Table 7.1). This changed the Chinese trade position from being a net importer for several decades to being a net exporter to the world. Due to the drastic output reduction in domestic production, resulting from three years of severe natural calamities and the man-made chaos of the Great Leap Forward, China shifted back to be a net importer again in 1961. During the 1960s, 43.3 million tonnes of grain were imported with the net trade of 27.7 million tonnes. This was an important factor in alleviating the damage caused by the big famine in the early 1960s. The trend of net imports continued until 1977, despite the Chinese government's efforts to achieve grain self-sufficiency in the first half of the 1970s.

The last phase started in 1978, characterized by an unprecedented record of net grain imports. From 1978 to 1983, total net imports registered 69 million tonnes, with average net imports of more than 12 million tonnes per annum despite the unprecedented annual growth rate of 4.9% in domestic grain production achieved during the same period. The unexpected change from the end of the 1970s had impacts on world grain trade which puzzled foreigners and caused increasing

Table 7.2 Grain imports and exports (in 10,000 tonnes)

Year	Total grain imports (a)	In which: Wheat imports (b)	% of the total (c)	Total grain exports (d)	In which: Rice exports (e)	% of the total (f)	Soybean exports (g)	% of the total (h)	Net trade column (d) – column(a) (i)
1953	1.457	1.36	93.3	182.62	56.12	30.7	86.47	47.3	+181.16
1957	16.68	4.99	29.9	209.26	52.95	25.3	114.11	54.5	+192.58
1962	492.30	353.56	71.8	103.09	45.79	44.6	25.92	25.1	–389.21
1965	640.52	460.45	71.9	241.65	98.49	40.8	65.32	27.0	–398.88
1970	535.96	277.04	51.7	211.91	127.95	60.4	47.01	22.2	–324.05
1975	373.50	349.12	93.5	280.61	162.96	58.1	40.47	14.4	–92.89
1978	883.25	766.73	86.8	187.72	143.52	76.5	11.29	6.0	–695.53
1983	1,343.51	1,101.91	82.0	196.31	56.59	28.8	33.39	17.0	–1,147.20
1950–59	83.54	28.29	33.9	2,228.10	708.95	31.8	1,088.86	48.9	+2,144.57
1960–69	4,924.86	4,328.51	87.9	2,155.20	993.03	46.1	603.47	28.0	–2,769.69
1970–77	4,280.46	3,672.85	85,8	2,142.70	1,222.72	57.1	307.49	14.4	–2,155.75
1978–83	7,898.13	6,497.23	82.3	962.14	521.10	54.2	112.91	11.7	–6,935.99

Sources: Columns (a), (d) and (i) are from *ZCWTZ 1952–1983*, p516
Column (b) is from *ZCWTZ 1952–1983*, p510
Columns (e) and (f) are from *ZCWTZ 1952–1983*, p496

concern amongst relevant international organizations. The motivation for China's grain trade is discussed next.

7.2.4 The motivation of China's grain trade

It is generally accepted that in the period analysed since the 1950s, China's grain imports and exports are based on a complex of several factors (Wong 1980, p118). There are four major ingredients contributing to the grain trade which have been pinpointed by Chinese authorities and researchers. These are discussed below.

The economic rationality of manipulating grain imports and exports Given the higher price of rice and lower price for wheat in the world market, an exchange of China's rice for foreign wheat is regarded by Chinese authorities and many researchers as economically rational (Timmer 1976, p66; Wong 1979, p454; Barnett 1981, p353; Timmer and Jones 1986, p154). As clearly pointed out by Mr Chen Ming (a high official in China's Ministry of Foreign Trade) in 1964, exporting rice, soybean and other processed food grain for the exchange of wheat "is a good means, ... of making money" (quoted from Wong 1980, p12). According to the statistics, the price for China's rice export averaged \$194 per tonne during 1953–83, while the price for wheat imports is \$93.5 on average for the same period (Appendix Table 1.6). The per unit price level for rice is therefore more than twice as much as that of wheat. Since China launched her industrialization programme under circumstances of very low per capita grain provision, the advantage of a neat calorific arbitrage to maximize foreign exchange earnings can be fully justified from such trade (even taking into account that the calorific content per unit of wheat is higher than that of rice).

The distributive motive As China remained a net grain importer since the 1960s, the distribution motive provides a further rationale for the significant net grain imports. Donnithorne (1970) argued that it was the limitations imposed by transportation facilities which accounted for much of the net grain imports in the 1960s as well as in the 1930s. Thus wheat is imported to supply urban consumers who are heavily concentrated on the eastern coastal area of three metropolitan cities (Beijing, Tianjin and Shanghai). The transportation argument has been echoed by Timmer (1976, p66), Walker (1977), Wong (1980, p125) and Kueh (1984).

The annual amount of wheat imports accounted for up to 20% of the total compulsory grain deliveries to the central government, Perkins (1975) thus argued that foreign grain imports were a method of reducing the pressure on the government's grain procurement programme. Indirectly they served to preserve the necessary incentive for peasant farmers. It also alleviated the possible conflicts with the regional authorities of grain surplus areas, who had every reason to keep stocks of grain on their hands for various purposes. This argument is strongly supported by Walker (1985) with his detailed systematic description of the fierce struggle created by the implementation of the compulsory quota system especially in the 1950s.

Shortage of domestic production Mah Feng-Hwa (1971) argued that food imports are primarily due to the inadequacy of domestic grain production. As a result, the wheat imports "provided the vital margin in keeping the food supply to the Chinese population above the subsistence level" (Mah Feng-Hwa 1971, p128). However, by pointing to the possibility of import substitution between grain and chemical fertilizer and the insignificant level of wheat imports as a proportion of domestic grain output, Wong (1980, pp122–123) rejected the idea that the "production shortages were the sole, or even dominant, explanation of China's wheat import policy". In Perkins' view, it would be "misleading or incorrect to explain China's long-term grain import policy in terms of domestic food shortages" (Perkins 1975).

Political considerations Political development is important not only in affecting the whole trade pattern in general (see section 7.2.1), but also grain trade itself. As pointed out by Wong (1980, p121), "while China was importing wheat at commercial prices, the bulk of her rice exports, especially in the 1950s and 1960s, was channelled through government-to-government sales, barter arrangements or even in the form of food aid, thus yielding relatively lower revenue". As a result, the terms of trade for rice and wheat of the nation were 1.79 and 1.70 respectively during 1961–69 and 1970–76, which is signficantly less than the world level of 2.17 and 1.82 during the same periods. This is primarily due to the political signficance of Vietnam and member countries of the Association of Southeast Asia Nations (ASEAN). During the 1950s, between one third to one half of Chinese grain exports, of which over half of the rice exports, went to the Soviet Union and Eastern Europe (Wong 1979, p452). Meanwhile about half of the wheat imports since the late 1970s came from the United States after the reconciliation in diplomatic relations between the two nations in the early 1970s (Barnett 1981, p355).

Although the general arguments about economic rationality, the distributive motive, production shortages and political determination are four major ingredients in influencing grain trade in the last three decades, the main themes for the periods of the 1950s, 1961–65, 1966–77 and 1978–83 can be further identified. During the 1950s, a large amount of agricultural products, especially grain products, were mobilized and exported mainly for the heavy imports of machinery and raw material to service the industrialization programme. As spelled out by Mr Chen Yun, the main architect of China's development programme in the 1950s, about two thirds of the 1.6 million tonnes of grain in 1953 was exported in exchange for machinery with Russia and other countries of the Eastern European Bloc, and another 17% for the exchange of rubber with Sri Lanka (Chen Yun 1953a, pp204, 205). China needs to export certain amounts of grain, oil-bearing products, meat and other agricultural products for it is absolutely essential for her industrialization programme. If she did not save agricultural products for export, there would be no foreign exchange for importing manufacturing facilities and industrial materials; and without importing large amounts of modern equipment, industrialization would be more difficult and slower (Chen Yun 1954, pp255, 256). Consequently, net grain exports amounted to 21.5 million tonnes (milled grain) in the 1950s (*ZCWTZ 1952–1983*, p516) notwithstanding the average per capita grain availability of only 250–300 kg (unmilled grain) during the same period (*ZCWTZ 1952–1983*, pp103, 158).

However, the serious shortage in domestic grain supply in the early 1960s forced the nation to slow down the pace of industrialization and to import large amounts of food grain for her people's survival. Due to the failure of the Great Leap Forward movement and continuing natural calamities during 1959–61, agricultural production collapsed. The gross value of agricultural output in 1961 was 30% less than that of 1957 (measured in 1952 price). An 18% net reduction in grain output was registered (see section 5.2.3). The per capita grain output was at the lowest of 216 kg (unmilled) in 1960. Consequently, famine loomed over the whole nation and the rural population decreased by more than 20 million people during 1958–60 (see section 5.2.3). In this situation, the pace of industrialization had to be set back and net import of grain products became very urgent and even of paramount importance (Chen Yun 1961, pp147,148,155). Thus, the composition of imports was adjusted considerably with the proportion of producer goods falling from about 90% in the 1950s down to around 60% during 1961–65 (*ZCWTZ 1952–1983*, p507), and a total of 21.6 million tonnes of net grain imports were registered during 1961–65 to help relieve the domestic food demand pressures (Appendix Table 1.15).

Since one period of political chaos was followed soon by another, i.e. from the Great Leap Forward (1958–60) to the Cultural Revolution (1966–77), the failure of "taking grain as the link" policy resulted in stagnation in domestic production. This in turn made continuing net grain imports the only way to fill the gap between production and utilization. During the era of the Cultural Revolution, although grain self-sufficiency was stressed throughout by the government, the combined effects of "squeeze agriculture hard" and "taking grain as the link" in the sacrifice of other components of the agricultural sector not only seriously harmed the development of cash crop production, but the per capita grain output also stagnated for 12 years between 1966–77 (section 5.2.3). Consequently, many peasant farmers were unable to provide their families with basic food needs (see section 6.3.1). For the nation as a whole, more than 10 million of the rural population were still suffering from the shortage in food grains (see section 2.5.2). Since there was no real political will to encourage domestic grain production, the only other solution was to remain a net grain importer.

Since 1978, the major drives for increased grain imports came from both production and consumption. In terms of production, deep lessons were learned by the Chinese government from the failure of agricultural development. One of them, as described by a governmental document, is "the adoption of policies and measures which were unable to arouse the peasants' enthusiasm for production" (from the "Decisions of Some Questions Concerning the Acceleration of Agricultural Development" which was adopted by the fourth Session of the 11th Central Committee of the Chinese Communist Party during 25–28 September 1979, quoted from Schran 1984, p432). On the production side, the main culprit was identified as the compulsory procurement scheme with its low purchasing price and high purchasing ratio. Therefore, in order to provide an incentive for grain production, not only should the purchasing price be raised (which was actually done in 1979 and afterward, see section 2.5.2), but also the high pressure for purchasing should be reduced. The reason is very simple, for without allowing a fair amount of food grain for self-subsistence, there is no room to encourage producers to produce more. The reduction of pressure to deliver official purchasing (that is, the reduction of the grain purchasing ratio) had to be compensated by increased net imports of grains from the world market, in order to provide grains for rural citizens other than grain growers. Another lesson from the historical development is the "inefficient implementation of the principle of all-round development of farming, forestry, animal husbandry, sideline occupations and fishery" (also quoted from Schran 1984, p432). This is a confession of the failure of the 'taking grain as the link' policy. However, encouraging the development of a diversified rural economy means more rural population would, in part or totally, retreat from grain growing, which in turn implies either a reduction in grain procurement or an increase in rationed grain.

The facts as far as consumption is concerned are, first, since the implementation of the new development policy in 1978, consumer's personal disposable income increased dramatically. For the whole nation on average, the per capita disposable income increased from 131 yuan in 1978 to 259 yuan in 1983 (in current prices, see Appendix Table 1.7), a total growth of 98%. After taking inflation into account, the real income level increased about 76% (in 1952 prices, see *ZCWTZ 1952–1983*, p370). Second, starting from such a low level of food consumption, a high pressure of food demand was created by the sharp increase in disposable income. According to Chen (1983b, pp13–14), total consumers' expenditure on food consumption of the nation was 101.8 billion yuan in 1981, which is 55% higher than that of 1978 (in current prices). In volume terms, the amount of food consumption increased more than 40% during the same period. To balance production with consumption, therefore required increased imports of grain products.

This completes the discussion of the evolution of China's grain trade and the political and economic environment in which it took place. The remainder of this chapter is focussed on an attempt to model the major grain trade flows.

7.3 Model Specification and Results

7.3.1 Specification of the grain trade function

Modelling China's grain trade is by no means an easy task. China is described by Timmer and Jones (1986) as "An Enigma in the World Grain Trade". An econometric analysis of rice exports and wheat imports from the 1960s to the 1970s was attempted by Wong (1980). He specified wheat imports in quantity as a function of: the unit price of imported wheat, domestic output of grain and wheat, and population growth. He tried formulations expressed in the current year or with a one year lag. For rice he suggested exports are determined by: the unit price of exported rice, grain output in quantity, wheat imports in quantity and population growth. Again, relationships were estimated using the current year and also one year lags (Wong, 1980, pp130–131). Unfortunately, there are problems with these specifications. The most critical is the omission of import substitution between wheat and fertilizer, and the influence of the distributive motive that is the effect of changes in domestic consumption conditions on grain trade. Consequently, most of the estimated coefficients were not significantly different from zero statistically, and the goodness of fit conditions of the models were extremely poor. Such results are not suitable either for structural analysis or prediction.

In this study, an econometric analysis of rice exports, wheat imports and net grain trade for the period 1961–83 was attempted. The theoretical model for China's grain trade specified that the quantity traded is determined by the price signals from the world markets, the domestic production and consumption balance, and the periodic shifts in the trade climate (which is mainly determined by internal and external political developments). The general function thus expresses the dependent variable, volume of rice exported or wheat imported as a function of the unit price of exported rice, the unit price of imported wheat, the unit price of imported chemical fertilizer, domestic grain output and consumption. Dummy variables were introduced for the periods 1966 to 1977 and for 1978 to 1983. Each of these is elaborated below following one further cautionary note. The separate analysis of production, consumption and trade is, strictly, only legitimate when the data series on each of these three are collated independently. Whilst the sources chosen purportedly satisfy this condition, some doubts remain, although the different disaggregation used for production (by commodity) and consumption (by usage) assists in ensuring statistical independence of the dependent variables.

The price effects Prices of the world market for rice exports, wheat and fertilizer imports were used as the major indicators of economic inducements to trade. The rice export price is a unit value of milled rice expressed free on board (f.o.b.). The wheat import price is a unit value expressed with costs of insurance and freight (c.i.f.), as is the price for fertilizer imports (see Appendix Table 1.16). Changes in prices are expected to have a direct impact on the trade quantity of corresponding products, either imported or exported. Thus, an increase in the world price of rice products may induce the nation to export more rice, an increase in world wheat price would also have a negative effect on the amount of wheat imported. From the point of view of the whole grain sector, given the foreign exchange constraint, an increase in price ratio between rice and wheat would be expected to encourage the country to export more rice and import more wheat. Changes in fertilizer price could influence the substitution between imports of fertilizer and wheat, and could mobilize more or less rice products for export.

Domestic grain balance The domestic balance between grain production and consumption is another critical factor influencing trade volume. First, fluctuations in production can change the nation's net trade position totally given the low level of per capita output. Evidence for this is the reversal of the net trade position in the 1950s and 1960s. Second, increased domestic consumption resulting from improvement in personal income can significantly push up the level of grain import. This is demonstrated by the accelerated grain imports since 1978 despite domestic production having increased at the same time. In summary, it is the combination of domestic production and consumption conditions which fundamentally determines the net trade position of China's grain sector. Although transportation is indeed an important factor influencing the nation's overall supply/demand balance, in reality, poor transportation has either increased the domestic redistribution cost which is primarily the concern of the Ministry of Grain rather than the Ministry of Foreign Trade, or it reflects immobility in domestic resources which directly manifests as supply shortages. As a result, it is not necessary to cope with this factor separately in modelling the national aggregate grain trade.

The impact of the general trade climate Like the policy effect on output response analysis in Chapter Four, the political climate is regarded as another determining factor on grain trade. The general trade pattern of the grain sector since the 1960s, especially wheat imports, can be divided into three different periods, from 1961 to 1965, from 1966 to 1977 and from 1978 to 1983. After the reduction in production caused by the Great Leap Forward movement, serious shortages in domestic grain supply forced the Chinese Government to import a considerable amount of grain from the world market, which changed the nation's position of net grain exporter to net importer. However, the Chinese Government changed its attitude from outward looking to inward looking during the ten years of Cultural Revolution starting in 1966. During this period, domestic self sufficiency in grain production was a constant theme of propaganda as a serious political issue for a socialist country like China. Grain imports were curtailed to the minimum level barely to cope with the minimum requirement for survival. The new development policy executed since 1978 changed the leaders' attitudes from inward to outward looking again. During the last few years, more grain imports have been regarded as vital both to encourage peasant farmers' production enthusiasm and to improve people's daily living standard (see section 7.2.4). Thus while 1966–77 is identified as a period of inward looking, both 1961–65 and 1978–83 are regarded as periods of outward looking, although the first was largely involuntarily and the last consciously. Dummy variables are employed to distinguish these three periods.

Finally, it is worth mentioning that whilst international agreements on trade can also be included as an independent variable in a trade function (see Simon 1980, pp15–24), as the treaties on China's grain imports and exports have never been publicly available they are not considered in this model specification.

7.3.2 The estimated results

7.3.2.1 Rice exports

The rice export function has the explanatory variables of domestic rice production (unmilled in units of 100,000 tonnes), domestic grain consumption (milled, in units of 100,000 tonnes, as a time-series of rice consumption is not available), the export price of rice products (US$/tonne), the price ratio between rice exports and wheat import, and of rice and fertilizer imports. The estimated results are presented in Table 7.3.

As shown by equations 7.1 –7.3, all three specifications of the production level

Table 7.3 Rice exports

Equation	7.1	7.2	7.3	7.4	7.5	7.6	7.7	7.8	7.9
Intercept	244.46***	259.49***	261.84***	83.66**	77.82**	101.92*	96.85*	59.84	35.84
	(4.18)	(4.65)	(4.89)	(1.61)	(1.75)	(1.37)	(1.62)	(1.29)	(0.73)
Rice output previous year	0.21***			0.66***	0.13**	0.17***	0.13**	0.13**	0.09*
	(3.25)			(2.89)	(2.42)	(2.82)	(2.42)	(2.48)	(1.50)
Rice output 2 year average		0.35***							
		(3.80)							
Rice output current year			0.60***						
			(4.11)						
Grain consumption	−0.25***	−0.36***	−0.55***	−0.19***	−0.13***	−0.21***	−0.15***	−0.15***	−0.11**
	(−3.28)	(−3.92)	(−4.30)	(−3.67)	(−2.78)	(−3.02)	(−2.35)	(−3.04)	(−1.84)
Rice export $/tonne	0.27**	0.29**	0.31**			0.06	0.07		
	(1.73)	(1.92)	(2.15)			(0.35)	(0.49)		
P.rice **exports** P.wheat imports				83.96***		78.72***		34.79	36.01
				(3.33)		(2.64)		(1.19)	(1.26)
P.rice **exports** P.fertilizer imports					31.34***		29.77***	24.09***	18.72**
					(4.38)		(3.73)	(2.58)	(1.87)
D_1 = 1 for 1966–1977									26.24*
									(1.33)
Price elasticity	0.52	0.56	0.60	1.58	1.02	1.59	1.11	1.44	1.29
R_2	0.30	0.38	0.42	0.49	0.60	0.47	0.58	0.61	0.65
F	4.09	5.45	6.32	8.03	11.92	5.78	8.64	9.49	8.27
D-W	1.46	1.64	1.764	1.24	1.86	1.29	1.90	1.70	1.60

Notes: Figures in the parentheses are the t-ratios of the estimated coefficients
***Significant at 1% level
** Significant at 5% level
* Significant at 10% level

(current year, lagged one year and the average of the two) produce significant coefficients). There are two explanations for this. On the one hand, the current year trade grain usually comes from the previous year's harvest. This is because exporting is not possible until after purchasing, processing (milling in this case), packing and mobilizing. On the other hand, harvesting conditions would be taken into consideration. Provided the current year's domestic supply is secured, a bumper harvest would encourage more exports and a poor harvest, less. The empirical evidence seems to suggest that the current year output is more relevant than the other judging from both the magnitudes of the coefficients estimated and the conventional statistical criteria. However, since the resource for the current year exports mainly comes from the previous year's harvest and trade agreements should also be taken into account in trade decision-making, rice output with a one year lag is therefore regarded as the main indicator for the relations between domestic production and trade. Apart from the production impact, evidence shows that rice exports are also significantly influenced by the domestic grain consumption and the export price.

Although a solid relationship has been found between the amount of rice exported and the rice price in the world market, prior knowledge suggests that the more important considerations are the ratios of prices of rice exports, wheat imports and fertilizer imports. This is supported by the evidence from equations 7.4–7.9. Equations 7.4 and 7.5 indicate that the two price ratios display significant coefficients when they are separately included. This is further supported by the improved goodness of fit as judged by the R^2 and F statistics and compared with these in equations 7.1–7.3. However, when more than two price signals are included in one equation, only one appears to be significant. This may be indicative of multicollinearity (see equations 7.6–7.9). It is suggested that the statistical insignificance of one of the pricing variables in equations 7.6–7.9 does not imply their irrelevance, but rather indicates the common movements among the three price variables, which has increased the standard deviations of the estimates.

Equation 7.9 is the unrestricted equation corresponding to the restricted equation 7.8. The validity of the restriction was tested with the F-ratio (see section 4.3.2.1) and the restriction was rejected at the 5% level of significance. The evidence from the F-test is echoed by the t-test on the estimated coefficient for the dummy variable for the period 1966–77 which was significant at the 10% level. This suggested that during the era of the Cultural Revolution more rice for export was extracted despite the stagnation in domestic consumption. This evidence was robust over a number of different model specifications.

In general terms, all the coefficients in equations presented display the expected signs. Positive signs for the coefficients of domestic rice production (all three specifications), rice exporting price, the price ratio of rice and wheat and between rice and fertilizer show that rice export responds positively to the changes in domestic production and the price signals of the world market. Meanwhile, rice exports respond negatively to the changes in domestic grain consumption. The positive relationship between the quantity of rice export and price ratio of rice and wheat implies that economic rationality was indeed an important determinant of rice trade. After taking major relevant factors into account; the nature of the resource endowment of Chinese agriculture (high rainfall and good irrigation conditions in south China, see Chapter One); the relative costs of rice and wheat production (the production cost per unit of rice is higher than that of wheat due to the intensive water and labour inputs); the relative calorific contents of rice and wheat (the calorific content of wheat per unit is higher than that of rice); and consumers' preferences (rice is the favourite food grain for most people especially for those in south China); a higher price ratio encourages the Chinese government to export more rice in order either to maximize her ability to import more wheat to maximize the calorific contribution or as a way of making foreign exchange for fuelling the whole development programme. By the same token, a higher price ratio between rice exports and fertilizer imports could induce the government to export rice in exchange for fertilizer, enabling even more domestic rice production and consequently more rice products for export. There is thus a multiplier effect from such trade circulation (the marginal productivity of fertilizer is around three in rice production under current circumstances, but as can be seen from Appendix Table 1.16, the price for rice per unit is about three to four times that of fertilizer).

The magnitudes of the estimated coefficients indicate that, a one million tonne increase in rice output induces only 9,000–21,000 tonnes of extra exports the following year, and 60,000 tonnes of additional exports in the current year. In terms of domestic consumption, one million tonnes more domestic grain consumption reduces exports only by 11,000–55,000 tonnes. This evidence confirms the prior knowledge that rice exports are a small proportion both in terms of domestic production and consumption. This explains why it is that whilst domestic production and consumption are vital in determining the changes in rice export, the reverse is not true. It provides further support for the idea of designating a recursive system to describe the internal relationships of China's grain economy. The own-price elasticities of rice export supply range from 0.52 to 1.59. The calculated price elasticities are thus very sensitive to the variation in model specification. The preferred equations 7.7 to 7.9 indicate a price elastic response of rice exports to the world price in the range 1.1 to 1.4.

The structural relationships discussed above are supported by the significance of the estimated coefficients; most of them have successfully passed the t-test at the 10% significance level or better (except as discussed already). However it has to be admitted that the adjusted coefficients of determination ($\bar{R}^2$) of around 60% for the preferred equations are somewhat lower indicating omitted variables.

Finally, the estimation was carried out using log-linear functional forms as well as the linear relationships discussed here. As the results were not materially different or better they are not presented.

7.3.2.2 Wheat imports

The model specification for wheat imports is fundamentally similar to that of rice exports. Current year total grain output and consumption are chosen as the indicators for the domestic balancing conditions of production and consumption.

There are two reasons for this. First, China has been a net grain importer since 1961, and grain imports, mainly wheat, are essentially taken to fill the gap between domestic production and consumption. The appropriate measure for this gap is thus based on aggregate grain production and consumption in the current year.

Second, since the limited imports of grain in China are aimed at servicing direct food grain consumption, especially for feeding the urban population of metropolitan cities of Beijing, Tianjing and Shanghai in the eastern coastal area (see section 7.2.2), the time lag for distribution is therefore small. So, unlike rice exports, wheat imports should relate to the current year's grain production.

Estimation was performed using both linear and log-linear functional forms. As the basic features of the model are quite similar to that of rice exports, only a few equations in each functional form are presented. The results, shown in Table 7.4, demonstrate that the importation of wheat is positively related to the changes in domestic consumption requirements. Given the domestic production level achieved, increased domestic demand means more wheat imported from the world market. Wheat imports are negatively related to the level of domestic grain production. This implies that, other things being equal, an increase in domestic production decreases reliance on foreign grain imports. Moreover, the quantity of wheat imported responds negatively to changes in the price of imported wheat, the price ratios of wheat and rice and between wheat and fertilizer prices. When the price ratio of wheat to rice is low, the trade pattern is to encourage rice exports and wheat imports in order to maximize economic benefit. On the other hand, when the ratio is high, rice exports and wheat imports tend to fall. The relationship between wheat and fertilizer importation is a matter of internal substitution. When the price for wheat imports is comparatively low, government would like to import more wheat directly instead of chemical fertilizer. However, the upward movement in price ratio of wheat and fertilizer induces the authorities to import more fertilizer to support domestic grain production and therefore reduce wheat imports.

Unfortunately, the empirical results do not provide very strong support for these hypotheses. Most of the coefficients of the price variables failed to pass the t-test. There are three possible explanations for this. First, while the wheat price in world markets was vulnerable to the changes in global supply and demand conditions as well as other shocks from the international economic system (for example, the dramatic price rise in the early 1970s due to the oil crisis and general commodity price crisis, see Appendix Table 1.15), the primary purpose of China's wheat imports was to fill the gap between domestic supply and demand. When this is further compounded by the bureaucratic style of trade manipulation, it is not so surprising to find that the sensitivity of China's wheat imports to the changes in world prices is considerably blunted. Second, as mentioned in section 7.2.2, China's wheat imports mainly came from the western countries. Wheat trade was therefore under strong influence from political and diplomatic developments. Although there has been a general trend of improvement in diplomatic relations between China and the Western society, trading conflicts with some individual countries occur from time to time. A typical example is the dramatic curtailment of grain imports from the United States in 1983 in retaliation for US restrictions of China's textile exports

Table 7.4 Wheat imports

	In linear form				**In log-linear form**			
Equation	7.10	7.11	7.12	7.13	7.14	7.15	7.16	7.17
Intercept	−767.14***	−488.98	−532.35***	−129.19	−6.16**	3.33	5.58**	−1.63
	(−3.90)	(−1.14)	(−3.15)	(−0.34)	(−1.99)	(−1.16)	(−2.07)	(−0.61)
Grain output	−1.60***	−1.31***	−1.17***	−1.14***	−6.65***	−5.42***	−3.61*	−3.59*
(100,000 tonnes unmilled)	(−4.00)	(−3.08)	(−3.44)	(−3.16)	(−2.92)	(−2.15)	(−1.59)	(−1.49
Grain consumption	3.38***	2.85***	2.68***	2.46***	8.81***	7.08***	5.67**	4.88**
(100,000 tonnes milled)	(4.64)	(4.00)	(4.41)	(4.01)	(3.28)	(2.55)	(2.19)	(1.83)
World wheat price	−0.43		−1.28		−0.11		−0.33	
(US$/tonne)	(−0.29)		(−1.07)		(−0.36)		(−1.22)	
P.wheat **imports**		153.75		−380.20		−0.27		−0.56
P.rice exports		(0.33)		(−0.87)		(−0.54)		(−1.21)
P.wheat **imports**		−146.34*		−21.21		−0.53		−0.19
P.fertilizer		(−1.48)		(−0.23)		(−1.31)		(−0.49)
D_1 = 1 for 1966–77			−212.34***	−213.92***			−0.41***	−0.37**
			(−3.54)	(−2.93)			(−2.73)	(−2.30)
Price elasticity	−0.07	−0.51	−0.21	−0.34	−0.11	−0.80	−0.33	−0.75
R_2	0.74	0.76	0.84	0.83	0.51	0.53	0.63	0.62
F	22.17	18.18	29.84	22.39	8.52	7.26	10.43	8.25
D-W	1.42	1.41	2.03	1.75	1.68	1.47	2.00	1.66

Notes: Figures in the parentheses are the t-ratios of the estimated coefficients
***Significant at 1% level
** Significant at 5% level
* Significant at 10% level

(Kueh 1984, p1270). This is another disturbance in the price response behaviour in wheat imports. Third, the blurred relations between price and quantity of wheat imports may also be affected by the uncovered effect of trade treaties, for frequently inertia can be created through agreements in trade.

Equations 7.10–7.13 were performed in linear function form and equations

7.14–7.17 in log-linear form. As shown by Table 7.4, a similar picture is displayed by the results with both functional forms. The adjusted coefficients of determination ($\bar{R}^2$) of the logarithmic equations show that between half to two thirds of the variations in the logarithm of wheat imports has been explained by the variations of the logarithms of the independent variables. This indicates that around two thirds to three quarters of the variations in quantity of wheat imports is associated with the variations of the independent variables. There are no significant differences in values of the D-W statistics between equations of the two function forms. Furthermore, F-tests of the validity of the restrictions between the pairs of equations 7.10 and 7.12, 7.11 and 7.13, 7.14 and 7.16 and between 7.15 and 7.17 are all rejected at the 5% level of significance. These are supported by the t-tests for the coefficients of the dummy variables for 1966–77. Three of them have passed at 1% (equations 7.12, 7.13 and 7.16) and another at the 5% level (equation 7.17). This evidence suggests that the level of wheat import during 1966–77 is significantly different from those other two periods. The finding is therefore consistent with the idea that the wheat imports suffered from the internal political turmoil of the Cultural Revolution and the external isolation during most of the period, and, consequently the implementation of inward looking policy with stress on the domestic grain self-sufficiency. For further use in simulation and projection, equation 7.13 is judged to be the best.

7.3.2.3 Net grain trade

Table 7.5 contains the results of applying the same analytical approaches to net grain trade. As China has been a net grain importer since the 1960s, the analysis concerns aggregate net grain imports. Equations 7.18 –7.21 show that net grain trade responds negatively to the domestic grain output. The coefficient indicates that an increase of 10 million tonnes in domestic production curtails net grain imports by 1.18 to 1.26 million tonnes. Net grain imports respond positively to changes in domestic consumption. The magnitude of this effect is that an extra 10 million tonnes of domestic consumption induces 2.48 to 2.83 million tonnes of extra net imports *ceteris paribus*. This implies that it is the increase in domestic consumption which has the largest effect on the level of net grain imports. The results once again confirm that Chinese decision-makers respond rationally to the changes in the world market price signals. The negative relation between the wheat importing price and net grain imports shows that, with other factors held constant, a rise in the market price for wheat reduces Chinese wheat imports, therefore reducing the scale of net grain imports. On the other hand, net grain imports respond positively to changes in rice price. This is consistent with the idea that an improvement in the rice market price can induce the government to export more rice in exchange for more wheat products thereby reducing domestic pressure on food demand, and consequently expanding the levels of grain imports and exports. This is further supported by the estimates of the coefficient on the price ratio between rice and wheat (see Equations 7.18 and 7.19). The positive relationship estimated confirms that the higher the price ratio between rice exporting and wheat importing, the better deal it becomes for the Chinese authorities to export rice in exchange for wheat.

Table 7.5 Net grain imports

Equation	7.18	7.19	7.20	7.21
Intercept	–800.85*** (–3.79)	–549.10** (–1.85)	–599.12*** (–2.55)	–383.65* (–1.34)
Grain output (100,000 tonnes unmilled)	–1.18*** (–3.03)	–1.20*** (–3.12)	–1.26*** (–3.11)	–1.25*** (–3.11)
Grain consumption (100,000 tonnes milled)	2.64*** (4.01)	2.48*** (3.72)	2.83*** (3.97)	2.58** (3.45)
Rice export price (US$/tonne)			1.46* (1.42)	1.17 (1.11)
Wheat import price (US$/tonne)			–3.79* (–1.36)	–2.66 (0.90)
P.rice exports P.wheat imports	119.08 (1.18)	91.61 (0.89)		
D_1= 1 for 1966–77	–429.03*** (–6.19)	–308.64** (–2.52)	–431.91*** (–6.14)	–313.71** (–2.38)
D_2= 1 for 1978–83		256.15 (1.19)		243.18 (1.06)
$\bar{R}^2$	0.89	0.89	0.88	0.88
F	43.35	35.76	34.29	28.96
D-W	2.53	2.75	2.58	2.75

Notes: Figures in the parentheses are the t-ratios of the estimated coefficients
***Significant at 1% level
**Significant at 5% level
*Significant at 10% level

Most of the regression coefficients passed the t-test. The high value of $\bar{R}^2$ suggests that about 90% of the variation in the net grain import has been explained by the models estimated. This is suppported by the F-test which is significant at the 1% level of significance and indicates that the model gives a good overall fit to the data. All this evidence bears out the correctness of the model specification.

Equations 7.18 and 7.20 are the restricted equations corresponding to the unrestricted equations of 7.19 and 7.21 respectively. The validity of the restriction in each case was tested using the F-ratio and accepted at the 5% level of significance. The restriction was therefore valid and the model accepted. The interpretation of these results is that while there is significant difference in net grain trade between 1961–65 and 1966–77, there is no strong evidence for the distinction between 1961–65 and 1978–83. This is consistent with the evidence from the separate rice exports and wheat imports analyses. The magnitudes of the coefficient on the dummy variable demonstrates that around 60–80% of the reduction in quantity of net grain import was caused by the policy stance adopted during the inward looking period of 1966–77.

7.4 Summary

This chapter has outlined the basic features of China's trade pattern. It is clear that the grain economy is not only an important part of the nation's agricultural production, but also the major component of her agricultural trade. The major components of grain trade are exports of rice and soybean and imports of wheat products. Because of the continuing reduction in soybean production, the importance of soybean exports have decreased. At the same time there has been an increase in rice exports. The southeast Asia region has always been the most important destination for China's grain exports, so was the Soviet Union and Eastern Europe especially during the 1950s. The United States, Canada, Australia and Argentina are the major sources of China's grain imports. There have been dramatic shifts in the Chinese net trade position of grain products during the last three decades. In the 1950s, China changed its position from being a net importer for several decades to be a net exporter. However, due to the failure in domestic production in the early 1960s, a large amount of grain was imported to alleviate the damage done by the big famine. Despite the attempt by the Chinese government to achieve grain self-sufficiency during the era of the Cultural Revolution (1966–77), the trend of net importation continued until 1977. Since the adoption of the new development policy started in 1978, an unprecedented level of net imports of grain products has been recorded.

The motivations for China's grain trade can be generally summarized as, (1) exploiting the principle of economic rationality; (2) the distributive motive; (3) dealing with shortages in domestic production; and (4) the political considerations. These were examined in detail for each of four sub-periods: the 1950s, 1960–65, 1966–77 and 1978–83.

During the 1950s large amounts of grain products were exported in exchange for heavy imports of machinery and raw materials to serve the ambitious industrialization programme. The massive import of grains during 1961–65 was primarily due to the collapse in domestic production which, in turn, was caused by the political movement of the Great Leap Forward. During the period of the Cultural Revolution, the failure of the "taking grain as the link" policy resulted in stagnation in domestic production. This necessitated continuing net imports of grain as the only way to fill the production deficit. Since 1978, an increase in grain imports has been consciously encouraged by the government for reasons concerning both production and consumption. The production motive itself involves two objectives. The first is to provide an incentive for peasant farmers to produce, by releasing some of the high pressure created by the demands of official grain purchasing. The second objective is to encourage specialized farming for better economic efficiency. On the consumption side, the aim is to improve grain availability to reduce the high pressure of demand for food grain which is created by the large increases in personal disposable income.

Quantitative analysis of these grain trading relationships were estimated for the period 1961 to 1983. The quantity of grain imports and exports was specified to be a function of the balance of domestic production and consumption, changes in price levels in the world market, and the shifts between inward and outward looking attitudes due to the internal and external political developments.

Models were estimated for rice exports, wheat imports and the net trade. For rice exports, results in Table 7.3 show that, the quantity of rice exported responded positively to fluctuations in domestic rice production (both in the current year and after a one year lag), to changes in the world market price of rice, the price ratio of rice and wheat and the price ratio between rice and fertilizer. The results implied that the greater domestic production, the more the rice products can be exported *ceteris paribus*. Also, the higher the price can be offered or the more favourable the price ratio between rice and wheat and of rice and fertilizer, the larger the volume of rice which would be exported. Changes in comparative price of rice and wheat enable benefits to be extracted by manipulating rice exports and wheat imports. Likewise, changes in the price ratio between rice and fertilizer dictate the economic rationality of importing more fertilizer to support domestic production enabling even more rice exports. Rice exports also respond negatively to changes in domestic grain consumption. The higher the domestic demand pressure, the lower the quantity which can be mobilized for export. In addition, there is evidence to support a significant increase of rice exports during 1966–77, over and above that dictated by the "normal" economic forces listed already. This resulted from the hard squeeze on agriculture exacted during the Cultural Revolution and the inward looking attitude to trade. It also serves to illustrate the critical importance of major rice importing countries to China's foreign relations when the nation was isolated from both the Western and Eastern Blocs.

Results for wheat import and net grain trade displayed similar features and correspondingly similar policy indications as for rice exports. As shown in Tables 7.4 and 7.5, both wheat and net grain imports respond positively to domestic grain consumption and negatively to domestic production. The responses of net grain imports to the price signals specified are identical to those of rice exports. For wheat imports, a negative relationship was found with the price ratio of wheat and fertilizer imports. This is consistent with the idea that imports of wheat and fertilizers may substitute for one another over a certain range. There is also strong evidence from both of these estimated functions that there were significant effects of the inward and outward looking trade policies pursued over the period analysed. Imports of wheat and consequently net trade of grain as a whole was curtailed considerably when the inward looking attitude was adopted. In general, all the functions estimated in this chapter demonstrate that in addition to the price signals of the world market and the shifts in trade policy, domestic production and consumption together worked as the key factors in determining China's grain trade.

Finally, it is worth mentioning that as in other research (see for example, Wong 1980), the analysis described here presumes that China is a price taker on world grain markets. There are two justifications for this. First, although China has been an important rice exporter since the 1950s, most of the trade was conducted under bilateral agreements and the price for rice exports was maintained below the market price for political reasons (see section 7.2.4). Second, although China imported considerable amounts of wheat during the early 1960s (1961–65) as well as since 1979, because of the large volume of total world trade in wheat products and the fact that Chinese imports were usually arranged "through long-term agreements designed to be stabilizing in the event of tight markets" (World Bank 1985b, p102), the effect of China's grain imports on the world market price is regarded as insignificant. These arguments are pursued further in the next chapter.

In summary, studies of the domestic food grain consumption and foreign trade in Chapters Five and Six have borne out the two hypotheses raised. That is, observed grain consumption levels are not only dependent on consumers' income level and selling price, but also, and indeed more importantly, on the production level achieved and the growth rate of population. It is also clear that production is one of the main factors determining grain demand, and supply and demand together work as key factors in determining grain trade. The basic nature of China's grain sector and the linkage between production, consumption and trade is thus clear.

So far, the historical development of Chinese grain production, consumption and trade have been scrutinized, theoretical hypotheses have been tested and questions about major policy issues answered. It remains to use these estimated models for intermediate and long-term projections and policy simulation. This is the subject of Chapter Eight.

Part III

Development Projection and Conclusions

Chapter Eight

Projections of Future Development

8.1 Introduction

Since it was announced in the late 1970s by the state that the key target for the first stage of China's modernization programme was to quadruple the gross value of industrial and agricultural output between 1980 and 2000, massive efforts have been made by Chinese professionals to study the future development of the agricultural economy, especially the grain sector. The most influential research published so far are: the *Study on the development of China's grain and cash crops* by the Chinese Academy of Agricultural Science (CAAS) (1985), *On strategic issues of China's agricultural development* by the Chinese Agricultural Zoning Committee (CAZC) (1985), and the projections of China's agricultural development to the year 2000 by the Ministry of Agriculture, Animal Husbandry and Fisheries (MAAF) which have been made available by the World Bank (1985b).

Following the analysis of the historical development, this chapter projects the possible future of China's grain production, domestic consumption and foreign trade to the end of this century. In doing so, heavy use is made of previous attempts to do this. Projections will be made for the individual products of rice, wheat, soybean and other grains and for total grain output. The four categories of domestic consumption, food, feed, seed and industrial use, are also projected independently. The projection of grain trade starts with the construction of a balance sheet which takes into account all categories of production and utilization. This is followed by discussions about the probable direction, scale and composition of grain trade to the end of the century.

The projections are presented for just three years, 1990, 1995 and 2000. For each year, projections are made for three alternative scenarios both for production and consumption. In the cases where no fundamental structural change is expected, the empirical models established in previous chapters are utilized in the projections. Otherwise, hypothetical projections are made based on the assumptions made for the explanatory variables. From the methodological point of view these projections are thus a combination of empirical prediction based on estimated equations and hypothetical projection. The chapter begins with a brief outline of the basic features of the second stage rural restructuring, which is regarded as the basic socio-economic environment for the future development of grain as well as the whole rural economy from the mid 1980s to the year 2000. This is followed by the projections of domestic grain consumption, production and trade. The main findings are summarized at the end of the chapter.

8.2 The Second Stage Rural Restructuring

The general principles for reforming China's economic management system are to combine market regulation with government intervention and co-ordination at the macro level; and to rely on economic levers, the law system and administrative means to steer the course of development rather than relying totally on the administrative system. Under the new economic mechanism, as described by the then party general secretary, Mr Zhao Ziyang in his main speech to the 13th National Congress of the Chinese Communist Party, "The state regulates the market, and the market guides enterprises" (*Beijing Review 1987*, p8). The aim of these reforms was to achieve both the mobilization of initiative amongst producers and the balanced development of the national economy. Given the sound results achieved from the first stage rural restructuring which introduced the household contract responsibility system at its core, the second stage, carried out by the government around the mid 1980s, aimed to consolidate these achievements. The three main policy changes were, an improvement in rural institutions and production organization, the abolition of exclusive rights of central government in the purchasing and selling of grain and other products, and the encouragement of all-round development of the economy. Each of these is elaborated below.

8.2.1 Further improvement in rural institutions and production organization

There are three main ingredients involved. First, as mentioned in Chapter Two, the People's Communes had been the grassroot government institutions in rural areas since the late 1950s. They wielded power in government administration and economic management. Since 1982, township governments have substituted for the communes and are responsible for administrative affairs and making plans for the development of the local economy and social infrastructure. The communes became pure economic organizations in charge of running co-operative enterprises. Peasant farmers were given greater freedom to determine their own production, although this was still within a framework of contracts made with local communities (Huan 1985, p16). Second, for the development of rural commodity production, the establishment of new types of co-operation organizations were encouraged. Since the early 1980s, there has been a fast growth in the number and activities of co-operatives. Examples are: regional co-operation, specialized production co-operation, co-operation in the integration of agricultural production, input supply, products processing and marketing services. These co-operatives take place between peasant households, between the collectives (either former production brigades or teams), and joint ventures between the collective, the state and the peasant farmers. All these co-operative organizations were formed voluntarily and developed on their own economic strength (Du 1985, p15). Third, with the spreading of the household contract responsibility system, peasant households specializing in particular types of production began to merge. These are the so-called "specialized households". Currently, more and more peasant farmers are willing to give up their land and devote themselves to other specialized production activities like animal raising, traditional handcrafts, small scale rural enterprises and other services. At the same time, others were enthusiastic to take additional land into cultivation. Within the farming sector, the division of labour was further intensified. In pursuing the aim of obtaining economic benefits from remunerative prices some peasants engaged mainly in grain production and others in the production of cash crops. According to Hagemann and Pestel (1987, p164), some 25 million peasant families, about one seventh of all rural households, had become specialized households at the end of 1984. The tendency of land to converge into the hands of the capable farmers has proven beneficial both to farming and to the development of other sectors of the rural economy (Luo 1985, pp145–149, 177). The combination of these three main features has fundamentally changed the traditional practice of "the three-level administration by the commune, the production brigade and the production team, with the production team as the basic accounting unit" system which had existed for more than two decades since the early 1960s.

8.2.2 The abolition of exclusive rights of central government in purchasing and selling of major agricultural products

Starting in 1985, the compulsory purchasing system for agricultural products in China was officially relinquished. It was replaced by the contracted purchasing system for grain and cotton products between the official marketing system and the producers, while other products were to be left to be regulated by the free market system (Chen 1985, p15). Under the new system, agricultural producers were no longer obliged to sell a portion of their harvests to the state. Instead, they were given greater latitude to decide what, how and how much to produce according to the contracts made with the official marketing system and the supply and demand conditions in the free market. The state participates in market regulation to protect the interests of both producers and consumers. In doing so, the government purchases 75–80 million tonnes of grain products every year through contracts made with the grain producers (which accounts for more than 50% of the compulsory purchasing of grains in 1984, see *TJNJ 1985,* p482), 30% of which would be bought at the quota purchasing price and 70% at the above-quota premium. Grain not included in the contracts can be sold on the free market, for which the quota purchasing price is taken as the minimum protective price by the state. In situations where the market price falls below the quota purchasing price, the government will buy the available amount of grain at the protected price (Zhao 1985, p17). Meanwhile, the grain rationing system was still in operation. Whilst there was no longer price distinction between purchasing and selling grain products in the rural areas, urban residents were still able to buy grain at the original official prices with coupons (Du 1985, p17). Controls on the purchasing price of pigs and selling price of pork were also relaxed and financial subsidies offered to the urban consumers to offset hikes in pork prices (Chen 1985, p15). Thus, whilst there were significant steps taken in the direction of liberalizing agricultural markets at the farm level, the extent of contract purchasing and the constraints imposed by the systems of input delivery still imposed some constraints and restricted the flexibility of producers.

8.2.3 Encouragement of the all-round development of the rural economy

In view of the scarcity of the arable land in China (see section 1.1.4), comprehensive development which emphasizes socialized, specialized and commercialized production was regarded by the government as a key policy issue for the further development of China's rural economy (Luo 1985, pp158–169). Reminded by the costly lessons from the pursuit of "taking grain as the link" policy, the developments of animal husbandry, forestry, fishery, as well as industry, commerce, transportation and other social services were all emphasized. This adjustment in rural economic structure was considered vital to assist transfer of surplus labour from agricultural production to non-agricultural sectors. This, in turn, would raise rural incomes and enlarge the market demand for agricultural products as a consequence. Currently, more than 20% of the 300 million rural labourers are engaged in non-agricultural occupations, the target set up by the government is over 50% by the end of the century (Du 1985, p17).

In addition to the three main areas mentioned above, policies such as increased government investment in and supply of industrial inputs for the agricultural sector; reforms of the banking system aimed at raising more funds for rural development; improvement in the commodity circulation system in rural areas; establishment of a comprehensive network system of agricultural scientific research, extension and training services; and attracting more foreign capital to fuel the development of the rural economy, were also undertaken by the government (see *People's Daily* 1986a, p3). Although the implementation of the second step rural restructuring outlined is a development process in itself, and detailed policy ingredients are subject to modification with changes in circumstances, it is clear that this is intended to be the general policy framework for China's agricultural and rural development through to the end of the century. Particular policy issues concerning the future development of grain production are examined in the context of the projection of grain production in section 8.4 below.

8.3 Projections of Domestic Consumption of Grains

The future consumption of grain products of the nation will be fundamentally determined by the growth rate of the economy and thus the improvement in per capita disposable income together with the pace of population growth. In launching its programme for modernization, targets were set to quadruple the gross value of industrial and agricultural output (GVIAO) between 1980 and 2000 (at an average annual rate of 7.2%), an increase per capita national income (PCNI) from about $300 to $800 (equivalent to a growth rate of 5% per annum) and to maintain a 1% natural growth rate in population (1.2 billion for the year 2000). These were set by the government as the major official targets for socio-economic development to the end of the century (World Bank 1985a). Given that annual growth rates of 9.1% and 6.8% have been achieved for GVIAO and PCNI respectively over the five years 1979–84 (in real terms) (see *TJNJ 1985,* p28; World Bank 1985a, p1), it is expected that these two targets can be achieved or exceeded. However, unlike the assumption made in the other research referred to above the official target of 1% population growth rate between 1980 and 2000 is regarded as unachievable (see section 8.3.1.1 below for detailed discussion).

8.3.1 Food grain consumption

Since the per capita direct human consumption of grain in China has reached a saturation point of 255 kg in the mid 1980s (*TJNJ 1987*, p711), instead of continuing to be constrained by the domestic production as was the case for the period of 1952–83 (see section 6.3.1), the future aggregate consumption for food grain will be essentially determined by the population growth and the level of per capita consumption. Consequently, the structural change from consumption growth limited by supplies to consumption growth dictated entirely by population and income growth makes it difficult to use the empirical models established in Chapter Six and made the hypothetical projection approach more appropriate.

8.3.1.1 Population growth

As mentioned above, the population for the year 2000 was targeted at 1.2 billion by the government, with an expected growth rate of 1% between 1980–2000. However, according to the official statistics, the actual population growth during 1980–85 registered 1.2% per annum (*TJNJ 1987*, p89). Moreover, according to the nation-wide sample survey carried out by the SSB, the birth rate of the nation in 1986 was 2.1%, after deducting the death rate of 0.67%, the natural growth rate was 1.4%. This made the total estimated population more than 1.06 billion for 1986, which shows a net increase of over 14 million from 1985 (*People's Daily* 1987a, p1). A 1987 sampling survey showed that the total population had reached a new record of 1.07 billion. It was estimated by the SSB that the natural growth rate for 1987 was likely to be around 1.48% (*People's Daily* 1986b, p1). There are several reasons which can explain this upward trend. First, bearing in mind that after the net reduction in population at the beginning of the 1960s, the nation was engaged in a high population growth period during 1962–73, the total population increased from 658.6 million in 1961 to 892.1 million in 1973, a net increase of 233.52 million and a registered annual growth rate of 2.6% for the whole period (see Appendix Table 1.8). As a result, "the arrival at marriageable age of China's Cultural Revolution baby boomers look like upsetting the planners' hoped for head count" (*The Economist* 1987b, p48). Second, since China is still a long way from nationwide welfare and pension schemes, for most of the rural population which accounts for nearly 80% of the total, the best expectation for a comfortable old age is to have more children, so there is great resistance to the "one child" policy. Third, since the implementation of the contracted responsibility system in rural areas, households have once again become the basic unit of production. Under this condition, more children in a household means more family labour in the near future and is consequently favoured by most peasant families. Currently, for those who have done well, the financial penalties for having more children are too small to affect a family, especially for those who can probably make extra profits by using another child in the fields. Fourth, for various reasons, the one child policy has never been enforced among China's national minorities. This, together with the considerable improvement achieved in daily living standards during the recent years, has led to minority nationalities registering a higher growth rate and thus a significant increase in population. According to the up-to-date sampling survey published by the SSB, the proportion of minority nationalities in total population has increased from 6.7% to 8.0% within five years between July 1st, 1982 and July 1st, 1987 (*People's Daily* 1987g, p1). (For detailed discussions of the effects of the introduction of the households responsibility system on population control, see Riskin (1987, pp302–304).) It is expected that most of these factors listed will continue to affect China's population growth during this century. Therefore, there are strong arguments to justify the suggestion that there is bound to be an overrun of the official target for population growth (NAZC 1985, p49; Qu 1987, p8). In this analysis, while 1% growth rate is assumed for the low projection scenario, 1.3% and 1.5% are chosen for the medium and high scenarios respectively. Based on the total population of 1045.32 million in 1985, the projected populations with alternative scenarios are presented in Table 8.1. It is believed that the medium scenario is the most probable outcome for the remainder of the century.

8.3.1.2 Per capita food grain consumption

The usual food consumption situation in China during the last three decades has been one of basic subsistence. Direct consumption of grains was the main source both for calorific and nutritional intake (see section 1.4.3). On average, 80% of the calorific consumption and 65% of the protein came from the direct consumption of food grain (Shen 1984, p117). However, per capita food grain consumption of the nation has reached over 250 kg per annum in the mid 1980s and this is generally regarded as the saturation point by most Chinese economists (CAAS 1985, pp381–

Table 8.1 Hypothetical projection of population growth for the years 1990, 1995 and 2000

Scenarios	Growth rate per annum (%)	Total population		
		1990	1995	2000
		-millions -		
Low	1.0	1,098.63	1,154.66	1,213.54
Medium	1.3	1,115.04	1,189.42	1,268.75
High	1.5	1,126.12	1,213.17	1,306.95

385). Like the consumer behaviour experienced in other countries, preference in food consumption will move away from direct consumption of grain and shift toward animal products, vegetable oils, fruit and processed food items, provided there is a continuing improvement in per capita disposable income and in grain availability (World Bank 1985b, p18). Consequently a reduction in per capita food grain consumption has been the prevailing assumption by several hypothetical projections, based on the expectation of improvement in consumption of other food items, especially that from animal sources (see Kueh 1984, pp1252–1266; Zhang Tong 1984, pp118–122; CAAS 1985, pp383–385; World Bank 1985b, pp18–20). The reduction in per capita direct grain consumption will be critically dependent upon the pace of improvement in per capita disposable income which will dictate the extent of substitution between food grain and the consumption of other food items. Given the structural change taking place in per capita food grain consumption it is not very sensible to use the positive income elasticity for direct grain consumption derived from the historical analysis of 1952–83 (see section 6.3.2.3) for projection purposes. Similarly, because they are based on the constraint of the "ideal" nutritional intake approach, the income elasticities for per capita food grain consumption during 1982–2000 of –0.16 – –0.33 which were suggested by the CAAS (1985, pp379–385) are not regarded as reliable. The preferred source of information on this matter is the systematic analysis of the possible changes in food consumption pattern to the year 2000 undertaken by Gong (1984, pp136–137). His figure of –0.06 is chosen as the income elasticity for the projection of future consumption of food grains in this analysis. Combining this with the per capita direct grain consumption of 256 kg in 1986 (*TJNJ 1987*, p711) and the assumption of 5% annual growth in per capita disposable income to the end of the century, the per capita food grain consumption is projected to be 237 kg in 1990, 232 kg in 1995 and 225 kg for the year 2000.

8.3.1.3 The projected total food grain consumption

With the assumptions made on population growth and per capita direct food grain consumption, the total food grain consumption of the nation for 1990, 1995 and 2000 are presented in Table 8.2. As shown by the alternative scenarios specified, the projected total food grain consumption in 1990 lies between 306.3 and 314.0 million tonnes and it is 321.2 to 346.0 million tonnes for the year 2000. All these figures are considerably higher than those projected by many Chinese professionals (see Gong 1984, Zhang 1984, CAAS 1985 and CAZC 1985). There are several reasons explaining such a significant difference. First, in projecting the future food grain consumption levels of the nation, everyone agreed that the level of total consumption requirement will be most critically determined by the growth in population. However, the official targets on population growth, which are 1.09 and 1.20 billion for the years of 1990 and 2000 respectively, were used by these researchers in projecting the future consumption of food grains. This is an under-estimation, for even if a 1% annual growth rate is applied for the period of 1985-2000 (which is the case for the low scenario in this analysis), the total population projected is still

Table 8.2 Hypothetical projection of total food grain consumption for the years 1990, 1995 and 2000

Scenarios	1990	1995	2000
	-million tonnes -		
Low	306.32	315.15	321.23
Medium	310.90	324.64	335.84
High	313.99	331.12	345.96

Note: Figures are derived by multiplying the total population projection in Table 8.2 with the correspondent per capita food grain consumption proposed. These are then multiplied by 1.15 for converting the milled to unmilled grain.

considerably higher than the governmental targets (see Table 8.1). Second, for those studies mentioned, projections of per capita food consumption were based on the concept of "objective" or "ideal" nutritional requirements. That is, a daily nutritional standard of 2,400 kcal and 70g of protein for an average adult as suggested by Shen (1984). It is reasonable to suppose that the pattern of future food consumption will be fundamentally determined by the consumers maximizing their utility rather than following advice either from nutritional scientists or medical authorities. The nutritional approach may be ideal but it is not necessarily a realistic description of consumers' behaviour. Third, previous projections of consumption guided by the proposed ideal nutritional standards, projected future per capita food grain consumption of less than 200 kg of milled grain. These projections have clearly misjudged the turning point of per capita food grain consumption. Although evidence suggests consumption levels of around 210–220 kg food grain during 1980–81, experience during the last few years suggests even higher levels of per capita food grain consumption while at the same time a considerable improvement in the consumption of other food items has also been achieved. Based on these arguments, it is believed that the level of total food grain consumption projected here is more reasonable than those made in the early 1980s. As the 1% annual rate of population growth from 1985–2000 is most unlikely for the reasons mentioned above and the 1.5% assumed in the high scenario would be regarded by the authorities as intolerable, the outcome from the medium scenario is regarded as most probable.

8.3.2 Projected feed grain consumption

Since the pattern of food consumption in China is under transition from basic subsistence to dietary improvement, consumption for feed grains will be another important component in future demand of grains. Equation 6.17 is chosen for projecting the level of feed grain consumption toward the end of the century, in which the estimated feed demand is determined by the consumption of pork and a time trend reflecting consumption of all other food items which include grain.

It is expected that due to the strong preference for pork consumption, this will continue to be the most important source of intake of animal nutrition. In turn, the future level of pork consumption will be determined by population growth and the improvement in personal disposable income. With population growth projected in Table 8.1 and the assumption of 5% annual growth in per capita disposable income within this century (see section 8.3.1.2), the remaining data requirement is an income elasticity of demand for pork. Three pieces of evidence were taken into consideration. First, for the last three decades between 1952 and 1983, while per capita disposable income increased from 97.4 to 258.7 yuan, an increase of 4.46 fold (see Appendix Table 1.7), during the same period, per capita pork consumption increased from 5.92 to 12.35 kg (see Appendix Table 1.13), a net increase of 109%. Therefore, for the whole period of 1952–83, the implied average income elasticity for pork consumption is around 0.24. Second, consistent with Engel's Law, there have been increases in income elasticity of demand for pork during the last three decades. For example, when per capita disposable income increased from 47.4 to 86.7 yuan during 1952–70, per capita consumption of pork registered a marginal growth of 0.1 kg (from 5.92 kg in 1952 to 6.02 kg in 1970), suggesting a very low income elasticity. However, given the improvement in personal disposable income from 91.3 to 258.7 yuan during 1971–83, the level of per capita pork consumption increased from 7.03 to 12.35 kg during the same period. This, very crudely, implies an income elasticity of 0.42. Third, for the Chinese populations outside mainland China, "the income elasticity of total grain consumption appears to have been about 0.25 over a considerable period of income growth" (World Bank 1985b, p18). With the expectation that the consumption patterns of mainland China towards the end of the century will probably follow the line of overseas Chinese, an income elasticity of demand for total grain of 0.25 was taken by the World Bank in projecting the per capita total grain consumption during 1980–2000 (World Bank 1985b, p19). Bearing all these in mind, values of 0.2, 0.3 and 0.4 are chosen as the income elasticities for pork consumption during 1986–2000 in this study for the low, medium and high projection scenarios respectively. Based on the nation's total pork consumption of 15.24 million tonnes in 1986 (*TJNJ 1987*, pp89,711) and the most probable population growth (medium scenario) projected in section 8.3.1.1, projections of total pork consumption for the years 1990, 1995 and 2000 are presented in Table 8.3.

According to the coefficient estimated using equation 6.17 on data for the period 1952–83, 2.38 kg feed grain (milled) is needed for a 1.0 kg increase in pork consumption. During the last three decades, pigs were predominantly raised by millions of peasant households as a family sideline. Each small scale production system with traditional technology undoubtedly involves a low level of technical efficiency, which is mainly manifested as a low conversion rate from feed to meat. However, such a dispersed small scale production system is capable of "minimizing the use of grain as animal feed and maximizing the use of by-products, households' wastes and low value feeds which have few alternative uses" (World Bank 1985b, p80). In

Table 8.3 Projection of the total pork consumption for the years 1990, 1995 and 2000

Scenarios	1990	1995	2000
	-million tonnes -		
Low	19.97	22.45	25.52
Medium	21.93	25.11	29.14
High	23.88	27.77	32.76

the future development of pig raising, while it is expected that the small scale household sideline production will remain the most important system for many years to come, more and more pigs will be raised by specialized households and large scale, commercialized pig raising companies (mainly in the suburbs of large cities). Such modern production systems can certainly improve technical efficiency in pig raising and reduce the feed conversion rate as a consequence. However, it would no longer be as effective in mobilizing and utilizing households' waste and roughages as small scale household production. Given these opposing influences it seemed reasonable to regard the coefficient derived from the historical analysis as appropriate for converting projections of pork consumption to feed grain demand.

Historically, the consumption of other animal products has been insignificant. For example, the average annual per capita consumption of fresh eggs was 1.0 kg in 1952 and increased to 1.9 kg in 1978; for that of poultry meat, it was 0.4 kg in 1952 and stagnated for 26 years until 1978 (see *ZCWTZ 1952–1983,* pp30, 50). Therefore a time trend was estimated for the overall increase in grain consumption for all meat production other than pork. The annual increase for 1952–83 was found to be 0.54 million tonnes on average. However, due to the substantial increase in personal disposable income and the strong drive for dietary improvement, production and consumption of these animal products have displayed an unprecedented growth since the late 1970s. Thus, the per capita consumption of fresh eggs increased from 1.9 kg in 1978 to 5.3 kg in 1986, and that for poultry from 0.4 kg to 1.7 kg at the same period (*People's Daily* 1987b, p1). It is expected that such a significant change will continue for the remainder of the century, therefore, much higher increase in feed grain than the historical record is expected for animal production other than pig raising. Incorporating this structural change, 1.5, 2.5 and 3 million tonnes of additional feed grain per year are added to the intercept for the low, medium and high scenarios respectively. These are additional to the continuing time trend estimated in equation 6.17. The resulting projections of feed grain demand are presented in Table 8.4.

8.3.3 Projected seed grain use

Equation 6.24 is used for the projection of future grain demand for seed. It will be recalled that the determining factors were the changes in grain growing area and a time trend. The seeding rate per hectare estimated in the equation was about 190 kg. This is regarded as appropriate for prediction purposes for neither significant

Table 8.4 Projection of the feed grain consumption for the year 1990, 1995 and 2000

Scenarios	1990	1995	2000
	(unmilled grain in million tonnes)		
Low	93.8	112.7	133.3
Medium	107.4	134.3	163.5
High	117.1	148.8	183.6

increase nor decrease in seeding rates is expected for the foreseeable future. The incremental change in seed requirement estimated as the coefficient of the time trend variable was 0.66 million per annum. This was explained by the significant changes which have occurred in cropping composition and also as a result of the introduction of HYVs in grain production (see section 6.3.4). However, there is no expectation of significant further adjustment in cropping pattern within this century (see section 8.4.1 and also CAAS 1985, p397). Moreover, the expected improvement in seed production and supply supervision would have the effect of reducing the observed average seeding rates. These factors could well offset the effects of further dissemination in HYVs (see section 8.4.1). Therefore, it is believed that there would no longer be a positive time trend from 1983. Using the assumption of most probable grain growing area of 112 million hectares for the remaining period of this century (see section 8.4.1), 29.5 million tonnes of unmilled grain is projected as the seed demand for all three years, 1990, 1995 and 2000.

8.3.4 Projected industrial use of grains

In the historical analysis, demand for grain for industrial use was classified into two general product categories. One was the grain requirement for making wine, biscuits and cake. Alcoholic drinks and processed foods are not part of the basic needs for people's daily life, the increase in demand is thus mainly driven by the improvement in personal disposable income. The other category includes all other industrial demand except for those in the first category, for example, ethyl alcohol, soysauce and vinegar making. These products are directly or indirectly linked to the basic needs of people's daily life, they are income inelastic and the increase in demand is essentially determined by the growth in population (see section 6.3.5). Since there is no fundamental structural change expected, equation 6.25 is used to project the future industrial use of grain products, thereby assuming that future grain demand for industrial use will continue to be determined by the total consumption of biscuits, cake and wine as well as a time trend.

During 1957–83, while personal disposable income of the nation increased 2.73 fold (see Appendix Table 1.7), per capita consumption of biscuits and cake increased 1.62 times (Appendix Table 1.11). The indicated average income elasticity is thus 0.59. The consumption of wine appears to be more income elastic than biscuits and cake. A 3.31 fold net increase in per capita consumption was recorded during 1957–83 (Appendix Table 1.11), with an implied income elasticity of 1.21 on average. Given the assumption of 5% annual improvement in per capita disposable income within this century (see section 8.3.2), and the income elasticities of 0.6 and 1.2 for biscuits and cake, and for wine respectively, this enables calculation of the industrial demand for grain for those products. In addition, it is expected that the historically estimated 64,100 tonnes annual increase in industrial requirements for grain will continue mainly due to the continuing growth in population. As a result, the projected industrial use of grain is 23.2 million tonnes for 1990, 29.6 million tonnes for 1995 and 38.1 million tonnes for the year 2000.

8.3.5 Discussion

The future grain consumption for food, feed, seed and industrial use projected above is summarized in Table 8.5.

A number of points can be made about these projections. First, as can be seen from Table 8.5, projected total grain consumption for the year 2000 lies between 522–597 million tonnes according to the alternative scenarios specified. This is substantially higher than those projected by Chinese researchers shown in Table 8.6. These indicate that the total grain consumption requirement was estimated at 480 million tonnes for the year 2000 by the CAZC (1985, p49); and 444–540 million tonnes was projected by the CAAS (1985, pp387–388). Given the 1983 total grain consumption of 367 million tonnes (Appendix Tables 1.5 and 1.10), the increase in grain consumption projected here from 1983 to the end of the century is between 156 and 231 million tonnes. This is an increase of 42–63% for the whole period. Compared with the historical records of a 2.5% annual increase during 1952–78, 7.7% for 1978–83, and 3.3% for the whole period of 1952–83 (all derived from Appendix Tables 1.5, 1.10), the average annual growth rate of 2.1–2.9% projected does not seem unreasonable. Meanwhile, according to the projected results in this study, the per capita grain consumption for the year 2000 is 430, 447 and 457 kg respectively for the three scenarios (Tables 8.1, 8.5). As the corresponding figure for 1983 is 359 kg (Appendix Tables 1.5, 1.8, 1.10), the annual growth rates implied are therefore 1.1%, 1.3% and 1.4% respectively; none of them is as high as the historical record of 1.5% during 1952–83. In addition, based on the assumption of 5% annual increase in per capita disposable income, the crude income elasticities implied for the total grain consumption for 1983–2000 on average are 0.22, 0.26 and 0.28 respectively. These elasticities are similar to the figure of 0.25 used by the World Bank (1985b, p19). It is suggested that the main merits of these projections compared to earlier work are first that particular attention has been paid to new information on the experience of recent years, especially for the recent trend of population growth and the current status of per capita food grain consumption. The other concerns the deficiency of taking the "ideal" nutritional approach in making grain consumption projections (see section 8.3.1.3), therefore, consumers' behaviour (reflected mainly by the income elasticity of demand for grains) has been one of the major considerations underlying these projections.

The second major point to emerge from Table 8.5 is that food and feed

Table 8.5 Projection of the domestic consumption of grains for the years 1990, 1995 and 2000

	1990			1995			2000		
	Low	Medium	High	Low	Medium	High	Low	Medium	High
	(unmilled grain in million tonnes)								
Food	306.3	310.9	314.0	315.2	324.6	331.1	321.2	335.8	346.0
Feed	93.8	107.4	117.1	112.7	134.3	148.8	133.3	163.5	183.6
Seed	29.5	29.5	29.5	29.5	29.5	29.5	29.5	29.5	29.5
Indus–trial use	23.2	23.2	23.2	29.6	29.6	29.6	38.1	38.1	38.1
Total	452.8	471.0	483.8	487.0	518.0	539.0	522.1	566.9	597.2

Table 8.6 A comparison of alternative projections of the nation's grain consumption for the years 1990, 2000

	1990		2000		
	Projected demand of this research (a)	Projected demand by CAAS (b)	Projected demand of this research (c)	Projected demand by CAAS (d)	Projected demand by CAZC (e)
	(unmilled grain in million tonnes)				
Food	306.3–314.0	256.7–294.6	321.2–346.0	228.2–285.0	260.0
Feed	93.8–117.1	94.9–112.2	133.3–183.6	145.4–184.7	150.0
Seed	29.5	21.8	29.5	20.0	
Indus–trial use	23.2	25.0	38.1	50.0	70.0
Total	452.8–483.8	398.4–453.6	522.1–597.2	443.6–539.7	480.0

Sources: Columns (a) and (c) are from Table 8.5 of this study
Columns (b) and (d) are from CAAS (1985), pp387–388
Column (e) is from NAZC (1985), p49

consumption will continue to be the two most important categories of grain demand in the future, accounting for over 88% of the projected total domestic grain consumption. This indicates that the future domestic requirements of grain are essentially determined by the growth in population and the pace of improvement in people's daily living standards. Since a higher population growth within this century than that targeted by the government is expected in this study, the amount of food grain and its proportion of the total domestic consumption projected are significantly higher than that indicated by other studies. For instance, the projected food grain consumption in this analysis makes up 65–68% of the total requirement for 1990, and 58–61% for the year 2000. These compare to figures of less than 65% and 51–53% respectively from CAAS and the World Bank arrives at the range 38–46% for the end of the century (1985b, p20). The feed grain consumption projected for the year 2000 indicates an annual growth rate between 6.7–8.7% during 1983–2000. This implies a substantial improvement in people's food consumption can be made. However, it must be pointed out that the proportion of nutritional intake derived from animal sources by the end of the century will still be lower than the current level in developed countries.

Third, the future composition of food and feed grain consumption will be critically determined by the growth in gross national product (GNP), in personal disposable income and that of population. For a given level of economic growth, the higher the growth in population, the lower the rise in per capita disposable income. Consequently, more of any additional grain is required as food grain and less as feed. On the other hand, the lower the population growth, the higher the improvement in personal income and, as a result, food grain, both in total amount and on per capita bases, will be reduced, perhaps substantially, while more grain would be available for animal production. Therefore, it is clear that the policy pursued by government in controlling the nation's population growth will be essential for the performance of future economic development as well as the improvement in people's living standards.

8.4 Projections for Grain Production

8.4.1 Basic assumptions for total grain production

It is expected that there will be no fundamental changes in the supply response mechanism of China's grain sector. This means that grain production to the end of the century will continue to be determined by the price of grain products, the shifts in policy pattern, the growing area of grain crops, the changes in weather conditions and the technological improvement in the grain sector.

8.4.1.1 Grain growing area

Official statistics on China's cultivated land was 98.4 million hectares in 1983 (World Bank 1985b, p28). Recently a nationwide agricultural resource survey carried out by the National Agricultural Zoning Committee reported that the actual cultivated area is around 130 million hectares, one third greater than the official figure (NAZC 1985, p52). Since the output figures are regarded as the most accurate in grain production statistics (see section 4.2.1), a larger cultivated area would simply mean considerably lower yields than the official figures. In projecting the future development of grain production, the official figures are used.

During 1981–85, the cultivated land of the nation registered a net reduction of 2.45 million hectares, about half a million per year, due to the shifts of marginal land from farming to grassland and afforestation, land losses to rural housing construction and urbanization and so on (*People's Daily* 1987c, p1). It is expected that a further reduction of at least 6.7 million hectares of cultivated land will occur within this century (NAZC 1985, p60). Because arable land has always been the most limiting factor for grain as well as other crop production, the government is determined to maintain current levels of cultivated area for the next 15 years by means of, first, tightening control on the loss of arable land from farming to other development purposes, and, second, by reclaiming new land for cultivation. At the same time, in response to the further development of irrigation and other measures for the improvement in production conditions, the multiple cropping index is projected to increase from the current level of 1.47 to 1.60 for the year 2000 (NAZC 1985, p60; WB 1985b, p29). In view of the small scope for further adjustment in farming composition between grain and cash crops production in the foreseeable future, a total grain growing area of 110–114 million hectares was assumed by NAAS (1985, p397) for the projection of grain production within this century. The figure assumed by the Planning Bureau of the MAAF was 113 million hectares to the end of the century (WB 1985b, p58). The figures used here for the grain growing area were 110, 112 and 114 million hectares for the low, medium and high projection scenarios respectively.

8.4.1.2 Prices of grain products

During 1952–84, the official composite average purchasing price for grain products increased 3.3% per annum (in current prices). The increases were 2.4%, 2.2% and 5% for rice, wheat and soybeans respectively (see Appendix Table 1.3). After taking into account the 1.1% annual increase in the national retail price index during the same period (*TJNJ 1985*, p530), the price for grain products increased 2.2% per annum in real terms, which was 1.3% for rice, 1.1% for wheat and 3.9% for soybeans.

It was made clear in Chapter Two that the grain market was a monopoly controlled by the government for most of the period from 1952 to 1984. With an imposed compulsory purchasing system, the official purchasing price worked as the paramount indicator of price movements. However, as mentioned in section 8.2, the compulsory purchasing system ended in 1984, and contracted purchasing has replaced the quota system for grain and cotton, leaving other agricultural products to be entirely determined by the market. Such a fundamental change makes the price signals for grain production in the future much more uncertain than before. In order to make an appropriate assessment about the possible price changes in the foreseeable future, it is worth discussing briefly the current situation of the grain market and the relevant issues involved. The contracted purchasing system has been in operation since the beginning of the summer harvest season of 1985. The state marketing system purchases a certain amount of grain annually and leaves the remainder to be regulated by the free market. The price for contracted purchasing is based on a weighted average of the quota and above-quota price. Generally the quota price had a weight of about 30% whilst the above-quota price made up the remaining 70%.

There were two main considerations behind these policy reforms. First, in the mid 1980s, the second stage rural restructuring was taking shape. In order to encourage the all-round development of the rural economy, to benefit further from the economics of scale and comparative advantage, it was considered by policy makers that reform of the grain purchasing system was vital to create a suitable climate to accelerate the whole process. Meanwhile, given China had six bumper harvests in a row and the total grain output reached an unprecedented record of over 400 million tonnes in 1984, a conviction grew amongst decision-makers that the grain supply situation had been fundamentally improved and the time was right to reform the old grain procurement system.

Second, as mentioned in Chapter Two, in order to encourage peasant farmers' enthusiasm for greater production, purchasing prices for grain and other major agricultural products were increased substantially since 1979. Given the government intention also of protecting urban consumers' interests, the inevitable consequence was a sharp rise in state financial subsidies. This is illustrated in Table 8.7. Since 1979 both the state budget spending on price subsidies in general and subsidies for grain and edible oils in particular increased rapidly. The price subsidies alone accounted for nearly 30% of the total state budget spending in the early 1980s. This was a major factor driving the state finances into continuing deficits. As can be seen from Table 8.7, the total budget deficits during 1979–84 amounted to more than 40 billion yuan. The situation was regarded as unbearable by policy makers in the mid 1980s. Measures for restraining the trend were therefore built into the reform of the grain purchasing system. In so doing, the price for contracted grain purchased was fixing using the 30:70 ratio. This was based on the fact that above-quota purchasing of grain had reached 70% of the total state purchases in 1984. This meant that starting in 1985, if producers sold more grain to the state than the 1984 level, they received exactly the same average price (based on the 30:70 ratio as mentioned) instead of the above-quota price as happened before. This implied that the marginal price which producers receive from selling extra grain was effectively reduced. For example, suppose the original quota price for 100 kg of grain was 100 yuan, the 50% premium for above-quota purchasing would make the price 150 yuan. This is the price grain producers could receive from additional selling to the state up until 1984. However, with the new weighted price system, the price producers would receive for an extra 100 kg sold to the state would be 135 yuan (0.3 x 100 yuan + 0.7 x 150 yuan). Simultaneously, the amount of contracted grain purchasing was also reduced. It was 75.6 million tonnes in 1985, 60.5 million tonnes in 1986 and was reduced further to 50 million tonnes in 1987 (Carter and Zhong 1988, p35).

Coincidentally, measures were taken to narrow the scope and therefore reduce the amount of grain which is directly rationed by the state. Industrial users of crop products in rural areas, for example in industries like textiles and brewing, restaurants and food manufacturers, who were all previously entitled to the planned grain supply were no longer within the scope of grain rationing from the official marketing system from 1985. Their requirements for grain had to be supplied either by the free market directly, or from the official marketing agency but charged to cover the full purchasing and marketing costs (Carter and Zhong 1988, pp47–48). These changes have thus narrowed the scope of the rationing system and consequently have reduced the quantity of planned grain supply. As a result, the state financial burden of price subsidies has been effectively reduced since 1985 (see Table 8.7). The market is playing a more and more important rôle in determining the nation's grain supply and demand conditions.

While the introduction of a contracted purchasing system for grains has facilitated the reduction of price subsidies, its negative effect on production has been equally clear. In 1985, the grain growing area fell by 4 million hectares from the previous year and output fell by 28 million tonnes. Several economic factors contributed to this reduction. First, as mentioned above, the contracted purchasing

Table 8.7 State budget income, spending and price subsidies*,1978-85

Year	(a) Total state budget income (billion yuan)	(b) Total state budget spending (billion yuan)	(c) Total price subsidies (billion yuan)	(d) % of budget spending	(e) Subsidies for grain and edible oils (billion yuan)	(f) % of budget spending
1978	112.11	111.10	9.39	8.45	3.63	3.27
1979	110.33	127.39	18.07	14.19	7.33	5.75
1980	108.52	121.27	24.21	19.96	10.80	8.91
1981	108.95	111.50	32.77	29.39	13.48	12.09
1982	112.40	115.33	31.84	27.60	14.72	12.76
1983	124.90	129.25	34.17	26.43	18.28	14.14
1984	150.19	154.64	37.00	23.93	20.50	13.26
1985	186.64	184.48	29.95	16.23	20.40	11.06

*Note: Price subsidies cover the price difference between products buying and selling and induced marketing costs. According to the Chinese State Statistical Bureau, there are three main items included in the price subsidies; viz, subsidies for domestic traded consumer goods; subsidies for input supplies for agricultural production; and subsidies for imported grain, cotton, sugar, chemical fertilizer and sprays.

Sources: Columns (a) and (b): *TJNJ 1987*, p617.
Column (c) : *TJNJ 1987*, p633.
Column (e) : from China's Statistical Information and Consultancy Service Centre, State Statistical Bureau.

system implemented in 1985 reduced the marginal price for extra grain sold. This was clearly a disincentive to grain producers. Second, the encouragement of all-round development of the rural economy successfully stimulated more non-grain and non-agricultural business in rural areas which quickly raised the opportunity costs of rural resources (land, labour and capital). Consequently, grain production became less and less attractive to the peasant farmers. Third, the continuing inflationary pressure since the early 1980s boosted market prices considerably both for consumer goods in general and industrial inputs for agricultural production in particular. This resulted in a sharp increase in production costs during recent years (see further discussion on this below). As a consequence, peasant farmers' enthusiasm for grain production diminished. Less land was used for grain and the total output fell in both 1985 and 1986. The government's responses to tackle this problem include: (i) a gradual reduction in the amount of contracted purchasing, enabling grain producers to sell more grain in the free market at a higher price and thereby increase their income; (ii) an increase in the grain purchasing price in 1986 and 1987 and provision of subsidized fertilizer and diesel oil to the producers (Carter and Zhong 1988, pp35, 107); (iii) increases in the government's spending on agricultural investment, both longer term investment in agricultural science and also more direct investment in capital; (iv) strengthening of the government's ability to monitor and control the shift of land between uses. These policy measures seemed to be working reasonably well in practice. Total grain output has gone up to the level of 400 million tonnes again in 1987.

It is now clear that under the new marketing system and pricing policy, several important factors must be taken into account in discussing the determination of future price signals for grain production.

The first is the comparative price between contracted purchasing and that in the free market. As mentioned in Section 8.2, the contract purchasing of grains will be maintained at a certain level per annum in the foreseeable future leaving the remainder to be regulated by the free market. Price movements in the free market will therefore have a great impact on the contracted purchasing price. The second concerns the price relationship between grain and grain using products. With the new system, prices for grain consuming products, such as meat, milk and eggs, are out of governmental control and determined by the market supply and demand conditions. Consequently, the market price fluctuation for these products will have a direct impact on the free market grain price as well as the price for contracted purchasing. The third factor concerns the price relationship between grain and other competitive crops. Bearing in mind that cotton is the main competitive crop to that of grain in the nation, the co-ordination of the prices for contracted purchasing of grain and cotton is vital for the central government. There is also competition at the regional level between individual grain crops and among grains and a wide range of cash crops. This is another complex dimension to be considered. The fourth relates to the price of grains and that for consumer goods in general and industrial inputs in particular. Price controls on industrially manufactured goods are also being lifted, in the long run this will boost the supply of industrial goods, but for the time being, it is mainly boosting the market prices. Consequently, "state price rises for agricultural produce in recent years have been mostly wiped out by the rapidly rising cost of manufactured goods sold in the countryside" (from the *People's Daily*, quoted from *The Economist* 1987a, p40). This is one of the main reasons that despite a substantial increase in grain purchasing price offered by the government just a few years ago, currently grain production remains the least attractive sector of the rural economy.

According to the experience of recent years, tremendous pressure has been created in pushing the grain price up due to the four considerations explained above. The extent to which further increase in grain price can be made in the foreseeable future is influenced by two other factors. One is the combination of a considerable increase in grain purchasing price since the end of the 1970s together with continuing bumper harvests which have increased government subsidy costs and created enormous pressure on the state budget. If the grain purchasing price continues to be substantially raised this will require more financial subsidies for urban consumption. The government has declared that no significant increase in grain purchasing price in the near future will be attempted and further improvement in producer's income level has to come from growth in production as well as efforts to reduce production costs.

The second concerns the price relation between domestic and international markets. Analysis has been made by the World Bank in which the above-quota price for major grain products in the nation were compared with border prices adjusted for transport, trade and processing margins back to the farmgate level. For brevity this could be called the farmgate border price. It was found that except for the above-quota price of rice, which was 80% of the farmgate border price, the above-quota prices for wheat, corn and soybean were 115%, 102% and 117% of the farmgate border price respectively. Thus they were all overpriced relative to the international prices (see World Bank 1985b, p43). Given the official exchange rate (2.8 yuan = $1) used in these calculations, there is obviously a problem of upward bias in the official exchange rate. Nevertheless, such indicative calculations provide evidence that the current grain purchasing price of the nation is not very far away from that of the international market. Moreover, it has also been projected by the World Bank that the world market prices for rice, wheat and corn will decline about 11%, 13% and 7% respectively in 1995 from the level in 1976–82 (in 1983 dollars) (see World Bank 1985b, p102 and Appendix Table 1.12). After taking all these factors into account, no dramatic increase in market price for grain products within this century is expected. For the projection, 0%, 1% and 2% annual price increase in real terms are chosen for the low, medium and high scenarios respectively.

8.4.1.3 Changes in policy patterns

Since the movement of the price ratio between grain and cotton coincided with the shifts in policy patterns during 1952–84 and this ratio was used as the indicator for the changes in policy patterns in the context of the grain supply response analysis (see Chapters Two and Four), it is once again used for the projection of the future development. It is suggested that in common with the periods 1952–57, 1961–65 and 1978–84, the general policy framework to the end of the century will be in favour of the future development of grain as well as agricultural economy. That is, there will be a continuation of the Policy Pattern A identified and tested in Chapters Three and Four. However, it is also important to be aware that the implementation of the policy measures involved in the second step rural restructuring is a process which will take time to be consummated. Moreover, there are several particular problems inherent in the future development of grain production. Unless all of them are properly settled the continuing growth in grain production could be substantially set back. Bearing all these in mind, the average price ratio of grain and cotton for the periods of 1956–60 and 1966–77 is taken as the input for the low projection scenario, that for the whole period of 1952–84 for the medium scenario, and the average of 1952–57, 1961–65 and 1978–84 was used in the high projection scenario.

8.4.1.4 Technological improvement

Despite the destruction which occurred during the Great Leap Forward and Cultural Revolution, tremendous efforts were made for the technological improvement of the agricultural sector during the last three decades (see Table 1.3). These efforts increased total grain output 2.9% per annum during 1952–84 despite a 9% net reduction in growing area which took place during the same period (section 1.3.2.2). It is clear that to achieve a sustainable growth in grain production toward the end of the century, further technological improvement is absolutely vital. In assessing the future potential for new technology, two factors are examined. One is the possibility of using more modern inputs, the other concerns the prospects of progress in agricultural science and technology, extension and training services and so on. In view of the detailed discussions on these issues presented by the World Bank (1985b, pp28–49) and Hussian (1986), only a brief outline is offered here.

The possibility of increased use of modern inputs It is well known that improved irrigation, better crop varieties and the utilization of chemical fertilizer have been three most important physical inputs for China's grain production over the last three decades. It will continue to be so for the remaining years of the century.

According to the NAZC (1985, p67), the current irrigated area in China totalled 40.5 million hectares (4.4 million hectares less than the official statistics), which accounts for about 40% of the total growing area and produces two thirds of the nation's grain output. In order to achieve the governmental target of 480 million tonnes of grain output for the year 2000, a 13.3 million hectare net expansion in irrigated area is needed. This would make the total irrigated area 53.8 million hectares by the end of the century. This plan has been confirmed by the MAAF which specifies the expansion of the irrigated area (World Bank 1985b, p33). The main efforts will be focussed on improving the utilization efficiency of the existing capacity. The target is to increase the water efficiency utilization coefficient from its current range of 0.3–0.5 to 0.6 within this century (NAZC 1985, p67). It is estimated that around 23 million hectares of the current cultivated area require drainage facilities to prevent waterlogging and government's projection for the year

2000 is 17 million hectares of more protected area. By doing this, it is hoped that the salinization problem in the North China Plain can be mostly overcome (World Bank 1985b, p34).

The application of chemical fertilizer in China's farming sector has registered an accelerated growth since the 1960s. By 1984, 17.4 million nutrient tonnes of chemical fertilizer was used at an application rate averaging 154 kg per sowing hectare (*TJNJ 1985*, pp252, 281). According to the MAAF's projection, the consumption of chemical fertilizer will reach 30 million nutrient tonnes by the end of the century, which is a 3.5% annual increase for the period of 1984–2000 (World Bank 1985b, p37). Organic fertilizers had long been the main sources of crop fertilization in China. In 1980, total nutrients from organics was estimated at the level of 14–21 million tonnes (World Bank 1985b, p35). If this is the case, it accounted for 52–62% of the total nutrients supplied in the same year (*TJNJ 1985*, p281). For the year 2000, 30 million tonnes of nutrients from organic sources is expected by the MAAF, which will make up half of the total nutrient supply for the end of the century.

The general situation for historical seed improvement has been investigated in section 6.3.4. For the year 2000, the governmental target is to produce up to 10 million tonnes of seed under NSC supervision. Meanwhile, the establishment of a nationwide seed testing system for certifying seed quality is also under consideration (World Bank 1985b, 40).

Improvement of agro-technical services New demands on agricultural extension services have been induced by the household contracted responsibility system. The future development of China's grain production will to a large extent depend on the success in providing small scale farms with the full range of technical services which are needed by the increasingly complicated production systems. A new programme for providing agro-technical services has been set up with the aims of further improving agricultural research and education and integrating all aspects of research, development and extension (World Bank 1985b, pp50–51).

First, to cope with the new situation created by the institutional reforms in the rural areas, a decision was taken to transform the existing county agricultural research institutes into agro-technical extension centres (ATECs) which combine the separate activities of extension, research, plant protection, seed production and fertilizer application services. Major functions of the ATECs cover training, demonstration and the provision of various services to producers. Contracts between the centres (or sub-centres) and individual households (or co-operatives) are encouraged. In doing so, bonuses will be received or penalties will be paid by technical service centres according to the achievement in production targets. Model households are also used as an important way of extending the improved technologies to producers. The ATECs are expected to be established in all the 2,300 counties of China by the mid 1990s (World Bank 1985b, p53).

Second, to accelerate the absorption process of new scientific achievements in grain and other production systems, the State Science and Technology Commission launched the "Spark Plan" in 1985. One of the important priorities of the Spark Plan is to improve ways of communicating research to the producer. The MAAF expects to see 70–80% of the research achievements to be put to work by the end of the century (Forestier-Walker 1987, pp41–42). Another massive programme called the "Bumper Harvest Programme" has been launched by the MAAF. According to the *People's Daily* (1987d, p4), 125 agricultural research institutes are involved in more than 1,000 technical assistance items around 300 production bases of grain and other major agricultural products. An increase of 1.6 million tonnes per annum of grain production is targeted from the implementation of the programme.

In summary, it is believed that the combination of massive modern inputs and close integration of agricultural training, demonstration and the provision of production inputs should be beneficial to the increasingly complex production systems in China's agriculture. It is therefore expected that the current time trend for technological improvement in grain sector can be continued to the end of this century.

8.4.1.5 Changes in weather conditions

Assessment of the possible impact from changes in weather conditions on grain production for more than one decade ahead is not possible. First of all, the long-term trend of global climatic changes is important but it is not yet clear from the meteorological research. From the agronomic point of view, on the one hand, continuing improvement in production conditions can weaken the effect of adverse weather on grain production. This is evident from the improvement in irrigation and drainage. On the other hand, the dissemination of HYVs, intensified use of chemical fertilizer and especially the increase in multiple cropping index could make the crops more vulnerable to the adverse natural conditions. In making projections of crop production the equations in Chapter Four were used. These require values for the weather variable which was specified as the expected grain growing area suffering from natural calamities. The value taken is the average proportion of growing area suffering from natural calamities (with more than 30% loss in yield) over the total planting area during the period 1980–84.

8.4.2 Assumptions for the production of rice, wheat and soybeans

Most of the assumptions discussed above are regarded as suitable for the separate projections of rice, wheat and soybean production. Therefore, the discussions below refer to distinctive features relating to the individual grain crops.

Rice In 1984, the area sown to rice was 33.2 million hectares. The future level of the rice growing area depends critically on the availability of well irrigated land. After taking into account both the inevitable loss of irrigated land to roads, urbanization and industrial development on the one hand and the possibility of bringing new irrigated land into cultivation on the other, the MAAF's Planning Bureau assumed that, for the remaining time period of the century, the rice growing area can be maintained at the present level. So 33 million hectares was used as the rice area in the World Bank projections to the year 2000 (World Bank 1985b, p58). Also, between 32 and 33.3 million hectares were chosen for the alternative projections of rice production for the years 1990 and 2000 by CAAS (1985, p397). The concensus of opinion is thus clear and so areas of 32, 33 and 34 million hectares are taken for the low, medium and high projection scenarios respectively in this study. In addition, continuing technological improvement is also expected in the future development of rice production. Currently, the expansion of the area planted to hybrids from 6.8 million hectares in 1983 to 13.3 million hectares in the year 2000 is assumed, which, according to the MAAF, would contribute nearly one third of the expected increment in rice production by 2000 (World Bank 1985b, p58). Detailed discussions about the possibility of and the measures to be taken to ensure future rice development can be seen in Mai (1984, pp3–11).

Wheat The general expectation for future wheat production by the Chinese government is more rapid output growth than other grain crops despite an assumed marginal decline in sown area. For the year 2000, the wheat growing area was targeted at 28 million hectares by the MAAF, which is 1.58 million hectares less than that in 1984 (World Bank 1985b, pp58–59; *TJNJ 1985*, p253). The assumption made by the CAAS (1985, p397) is quite similar, for 26.67 to 28 million hectares were set for the projections of the future development of wheat production. Meanwhile, a wheat growing area of 28 to 28.7 million hectares was used by Yan (1984, pp12–26) for the output projections to 1990 and 2000 who provided detailed analysis of both production and consumption issues. Mainly due to the difference in expectation of future population growth (see section 8.3.1.1), 28, 29 and 30 million hectares of wheat growing area are assumed for the alternative projections in this analysis.

Soybean For the reasons mentioned in Chapter Four (see section 4.3.2.4) soybean production was discriminated against for most of the last three decades. This resulted in a considerable decline in sown area and stagnation in yield. However, soybean production has recovered substantially since the end of the 1970s for several reasons. One is the significant improvement in per capita grain availability which has, to a large extent, released the pressure on rural grain self-subsistence. This in turn gives the opportunity to peasant farmers to grow soybean, which is low in yield but high in nutritional content. The second is that an increase of more than 50% in soybean purchasing price was offered during 1978–84 (see Appendix Table 1.3). This made production more profitable than ever before. The third is that the 1:1 ratio (between soybeans and grain) in soybean procurement was replaced by 1:2 (one tonne submission of soybean may count as two tonnes fulfillment of grain) since the end of the 1970s. This was another factor which substantially changed producers' attitudes to soybean production. As a result, the soybean growing area increased from 7.14 to 8.30 million hectares and output increased from 7.57 to 11.61 million tonnes (*TJNJ 1987,* pp164, 170). Given the importance of soybean production in improving people's dietary and nutritional intakes, it is both in the government and producers' interests further to boost the production in the foreseeable future. Having taken all these factors into account, several distinctive measures are taken for the projections of the future soybean development. First, 1%, 2% and 3% annual real price improvement are assumed for the low, medium and high projection scenarios respectively. Second, the average price ratio between soybean and cotton for the whole period of 1952–84 is taken as the indicator of the policy pattern for the low projection scenario, and the average value of this ratio for the periods of 1952–58, 1961–65 and 1978–84 is chosen for both the medium and high scenarios. Third, a continuing increase in soybean growing area is expected, and 9, 10 and 11 million hectares of growing area are assumed for the three scenarios respectively. Fourth, because of the structural change taking place since the end of the 1970s, a time trend in technological improvement on soybean production is expected and the average annual production increase due to the improvement in yield during 1978–86 is assumed to apply to future soybean production. Consequently, in using equation 4.29 for projection (see Table 4.8), an additional 0.35 million tonnes of annual output increase is added in the intercept to cope with structural change. Further detailed analysis about the future soybean development can be seen from Zhang (1984, pp48–60).

To ensure consistency, no attempt is made independently to project the future development of other grains. Instead, the difference between projected total grain output and the projected outputs of the individual crops of rice, wheat and soybean is taken as the projected production level for other grain products.

The equations used for the projections of total grain, rice, wheat and soybean are equations 4.9, 4.15, 4.22 and 4.29 respectively. A summary of all the assumptions made in the projections of crop production to the year 2000 is shown in Table 8.8.

8.4.3 Results: presentation and discussion

Putting all the assumptions into the estimated supply equations, the projected grain outputs for the years 1990, 1995 and 2000 are shown in Table 8.9. The projected total grain output for the year 2000 ranges from about 500 to 585 million tonnes according to the alternative scenarios specified. Compared with the actual production level achieved in 1985, around 120–206 million tonnes of net increase in grain production is projected within 15 years. These require an annual growth rate of 1.9%, 2.4% and 2.9% for the low, medium and high projection scenarios respectively. A comparison of projected grain outputs between this study and that made by CAAS and MAAF in the early 1980s is presented in Table 8.10. Generally speaking, the projected output levels in this study are higher and show a wider range of variation. This is partly due to the differences in assumptions made, for example, the expected growing area of grain crops (see sections 8.4.1 and 8.4.2), and partly because of differences in the method of projection.

In the projections conducted by Chinese professionals in the early 1980s, three approaches were taken. One was the production function approach, in which the historical relations between production inputs and growth in yield were scrutinized, and marginal productivity of major inputs, such as chemical fertilizer, irrigation and high-yielding varieties was derived. The prospects of yield improvement were

Table 8.8 Assumptions for the projections of China's grain production to the year 2000

	Annual price improvement for grain products (%)			Differentiations in Policy Pattern (using price ratios of grains and cotton as the indicators)			Grain growing area (million hectares)			Weather conditions	Technological improvement
	Low	Medium	High	Low	Medium	High	Low	Medium	High		
Total grain	0%	1%	2%	Average of 1958–60 & 1966–77	Average the whole period of 1952–84	Average 1952–57, 1961–65 & 1978–84	110	112	113.3	The average proportion of growing area suffering from natural calamities during 1980 multiplied by the future growing areas proposed.	Continued
In which:											
Rice							32	33	34		
Wheat							28	29	30		
Soybean	1%	2%	3%	Average the whole period of 1952–84	Average 1952–57, 1961–65 and 1978–84	Average 1952–57, 1961–65 and 1978–84	9	10	11		Annual increase of 0.35 million tonnes

Source: see text.

Table 8.9 Projected grain outputs for the years of 1990, 1995 and (unmilled grain in million tonnes)

	1990			1995			2000		
	Low	Medium	High	Low	Medium	High	Low	Medium	High
Total grain	417.3	444.8	468.7	458.5	493.6	526.2	499.7	542.7	585.4
In which:									
Rice	169.6	184.0	197.0	181.7	198.6	214.5	193.9	213.3	232.6
Wheat	71.4	81.2	90.1	77.3	88.7	99.5	83.3	96.2	134.8
Soybean	13.5	15.4	16.2	14.8	17.8	19.1	16.2	20.3	22.1
Other grains	162.8	164.2	165.3	184.7	188.5	193.2	206.4	212.9	196.0

Notes: Figures for other grains are derived by subtracting the projected outputs for rice, wheat and soybean from the corresponding figures for total grain output.

Table 8.10 A comparison of alternative projections of grain output for the years 1990 and 2000

	1990		2000		
Source of projection	This study (a)	CAAS (b)	This study (c)	CAAS (d)	MAAF (e)
	(unmilled grain in million tonnes)				
Rice	169.6-197.0	187.5	193.9-232.6	200.0-212.5	208.0
Wheat	71.4-90.1	86.0-90.3	83.3-134.8	100.0-105.0	99.0
Soybean	13.5-16.2	15.4	16.2-22.1	19.3-20.7	} 173.0
Other grains	162.8-165.3	150.8-152.3	196.0-212.9	171.5-188.7	
Total	417.3-468.7	439.7-445.5	499.7-585.4	490.8-526.9	480.0

Sources: Columns (a) and (c) are from Table 8.9 of this study
Columns (b) and (d) are from CAAS (1985), p397
Column (e) is from World Bank (1985b, p58)

therefore based on the presumption of a continued increase in major inputs. The second was based on a regional analysis. In doing so, grain production of the nation was disaggregated into low-, medium- and high-yielding regions. These generally coincide with the west, central and east of China respectively. The prospects of future yield improvement for different regions were studied and the time periods for the low-yielding region (west China) to catch up with the medium-yielding region (central China) and the medium region to the high-yielding region (east China) were estimated. The weighted crop yields for the nation were then derived and were used in projecting the future development of China's grain production. The two approaches mentioned have always been used by the MAAF in projecting the intermediate and long-term development of China's agriculture. Another approach, which is basically an econometric approach, was also tried by the CAAS. In doing that, time-series data on grain production (output, growing area and yield) for more than 2,000 production units (county or county equivalent) were collected. A linear relation curve between crop yield and time trend was estimated for individual units using ordinary least squares. The curve derived was then used to project the future grain yield of the corresponding production unit subject to the constraint of the crop growth curve (the logistic curve). A weighted crop yield of the nation was thus derived with aggregation. Such a highly disaggregated approach created a huge data processing and calculation task, but the result was not considered to be very worthwhile for it proved impossible to estimate an explicit production relation (either technical or economic) due to the lack of detailed information.

Generally speaking, these three approaches are all concerned with the technical relations of grain production, either explicitly or implicitly. No economic factors were directly considered in conducting these projections. The production projection made in this study is based on the supply response functions derived from the historical development analysis. While the projection was directly targeted on the national aggregate level, key policy issues, especially the price for grain products, were directly taken into consideration in making the projection. Bearing in mind that China's economic management system is under a transition from the highly controlled system to the greater use of the market mechanism, market forces, especially price effects, will play a more important rôle in determining the future development of the nation's grain production. It is for this reason that the production projections conducted in this study may prove to be valuable and novel.

Two issues emerge from examining the projected output in Table 8.9. One is the necessity for such a substantial growth in production. The other concerns the possibility of achieving a sustained production growth toward the end of the century and the most probable scenario to be expected. The necessity for the growth in output has been justified in detail in section 8.3 in which a significant increase in consumption was projected. The possibility of such growth is examined below. A generally favourable social economic climate for the future development of grain as well as the whole rural economy is being created, and this is essential for the success or failure of the future development of grain as well as the whole rural economy. However, it is acknowledged that the implementation of policy measures covered by the second step rural restructuring and the establishment of "market-regulated economic mechanisms formed under the guidance of the socialist planning system" is indeed a slow process and subject to set-backs. Within this general framework, sustained growth in future grain production will be critically determined by three particular issues.

The first concerns the maintenance of the requisite amount of arable land needed for grain production. Given the significant loss of arable land to non-agricultural use on the one hand and the considerable shift in farming composition from grain to cash crop production since the end of the 1970s (see Table 1.6, sections 1.3.2.1 & 8.4.2.2), tough policy measures may be required to secure the amount of land required for grain production. In so doing, tight control on the shift in land use from farming to rural housing and industrial use will be absolutely necessary. Currently, new regulations have been established and responsible institutions have been set up inside the administrative system. Meanwhile, investment must be made available for reclaiming new land for cultivation to compensate for the inevitable loss of cultivated land. Furthermore, co-ordinating the profitability between grain and cash crop production must be taken seriously in making governmental pricing policy in order to prevent undesirable changes in the composition of output. In addition, attention is necessary to encourage agronomic efforts aimed at increasing the multiple cropping index. It is believed that any failure in these areas will threaten the maintenance of the current level of grain growing area through the next decade.

Second, given that there is little chance for increasing growing area in the foreseeable future, yield improvement will be the principal source for the further growth in grain production. As can be easily derived from Table 8.6 and 8.7 the average yield of grain production is 4.54 tonnes/ha in the year 2000 with the low projection scenario, and 4.85 and 5.17 tonnes/ha for the medium and high scenarios respectively. Compared with the current yield level of 3.48 tonnes/ha in 1985, annual growth rates of 1.8%, 2.3% and 2.7% will be necessary in the low, medium and high scenarios respectively. In achieving this, massive physical inputs and tremendous efforts in scientific study and extension service will have to be made (see section 8.4.1.4). These will not be possible without massive productive investment.

Before the implementation of the household contracted responsibility system, the government and the rural collectives (including communes, production brigades and teams) were the main sources of agricultural investment, and the rural collectives contributed about 60% of the total mainly by means of labour mobilization for capital projects (Hussian 1986, p9). Since the beginning of the 1980s, government direct investment in agriculture (for the definition and coverage of governmental budgetary investment in agriculture, see Lardy 1983, pp128–138) was cut and that from local collectives decreased also. The former was due to the constraint on budget spending. In the case of the latter, the reduction in resources available was caused by institutional change. For the foreseeable future, agricultural investment by rural collectives is expected to continue to decline and that from government sources to recover but with a smaller proportion directed towards crop production (World Bank 1985b, pp46–50).

The total investment requirement in the agricultural sector (including capital and recurrent investments) between 1981–2000 was estimated at 1,422 billion yuan by the CAAS (1985, pp424–425), in which 1,010 billion yuan was for agricultural production and 412 billion yuan for rural industrial development. In mobilizing all of the needed funds, it is expected that 6% will come from government budget spending on agriculture, 12% will be transferred from profits made in rural industrial sector, 15% will be financed by the rural credit lending, and the remaining two thirds should be contributed by private savings for productive investment. It is quite clear that without such massive productive investment in the sector, it will be impossible to sustain agricultural growth at the rates suggested in the above projections. The challenge for the central government is not only to secure this budget spending on agriculture, but also, and indeed much more importantly, on how to induce the enthusiasm of lower levels, i.e. provincial and local finances, and especially the 800 million rural population, to invest in agriculture. Currently, several policy measures have been taken to encourage agricultural investment. For example, land contracts to peasant households which were originally limited to two or three years only have been extended to 15 years or even longer since 1984 (Hagemann and Pestel 1987, p164); rural credit institutions have been handed back to the co-operatives and interest rates of the State Agricultural Bank manipulated in favour of productive investment in agriculture (*People's Daily* 1987e, p1). Joint investments between central and local government for the construction of grain production bases are pursued (World Bank 1985b, pp30–31; *People's Daily* 1986b, p1). All these measures are taken primarily to induce savings of the lower levels which are growing most rapidly to be invested in grain as well as other agricultural production. It is clear that more effort has to be made in this respect if the historical tragedy of neglecting agriculture is to be avoided.

The last critical issue concerns a sound and sensible pricing policy. With the introduction of the market mechanism and abolition of a centralized controlling system, producers will enjoy a greater degree of freedom in decision-making and production management. In this situation, state price manipulation will become simpler and it could have an unprecedented impact on grain production. Without getting the price right and thus making agricultural production attractive, the aims of maintaining the current level of sowing area, mobilizing massive productive investment in agriculture and all the other relevant efforts are doomed to fail. As grain production is generally the least profitable sector in rural areas, especially for the coastal areas which have a better social infrastructure for the development of specialized and commercially orientated activities (see sections 5.1.2 & 8.4.1.2), inducing the local government and basic accounting units to protect grain production by means of transferring profits from other sectors for subsidizing the grain sector is becoming a key policy issue for the central government (see *People's Daily* 1986a, p3).

In summary, the general socio-economic climate pictured by the second stage rural restructuring is regarded as a necessary but not a sufficient condition for the desirable future development of grain production. Particular issues of securing arable land for grain growing, raising massive funds for productive investment in the sector, and getting the price right are the three most important policy issues for a sustained growth in grain production within this century to meet the challenge of the expected huge demand pressure. It is fully realized that successful achievement in all three major areas mentioned is by no means easy and tremendous efforts need to be further pursued. However, it has also been clear that, since grain is such a vital product for the nation (a bitter political dispute was caused inside the central government by a marginal reduction in growing area and production level between 1984–85), vigorous efforts will be made wherever and whenever it is justified. Bearing all these considerations in mind, it is believed that China's grain production will achieve a substantial growth during the next 15 years and the medium scenario projected is regarded as the most probable outcome.

Chapter Nine

Outlook for Trade and World Market Impacts

9.1 The Current Situation of China's Grain Imports and Exports

Chapter Seven analysed in detail China's net trade in grain over the period 1952–84. In this section attention is focussed on more recent developments in grain trade.

The official statistics of Chinese grain imports and exports during 1983–86 are presented in Tables 9.1, 9.2 and 9.3. Figures in Table 9.1 are based on customs data which therefore indicate the annual flow of grains in and out of the nation's territory. Those in Tables 9.2 and 9.3 are from the Ministry of Foreign Economic Relations and Trade (hereafter MOFERT) which refers to grain trade conducted by the foreign trade department. Data from both sources include entrepôt trade of grain products which has grown rapidly during recent years. Several important points are revealed by these statistics.

First, the general trend of China's grain trade since the late 1970s has been reversed during the mid 1980s. As discussed in Chapter Seven, until the late 1970s China's grain supply and demand was very tightly balanced. With the implementation of new agricultural development policy at the end of 1978, more grains were imported to improve grain supply conditions and release high domestic pressures. Therefore, grain exports were reduced from 1.9 million tonnes in 1978 to 1.3 million tonnes in 1982. During the same period, grain imports were sharply increased from 8.8 million tonnes to the record level of 16.1 million (see *TJNJ 1983*, pp422, 438). Consequently, the net import of grains increased from 6.9 million tonnes in 1978 to the unprecedented level of 14.8 million tonnes in 1982 (see Appendix Table 1.15).

The great achievement of the new policy has been the boost to domestic grain production since the end of the 1970s and thereby the significant improvement in domestic grain availability. Trade statistics from the customs show that the nation's grain exports increased from 1.1 million tonnes in 1983 to 9.4 million tonnes in 1986. At the same time, total grain imports decreased from 13.5 million tonnes in 1983 to a low of 6 million tonnes in 1985 before recovering to 7.7 million tonnes in 1986 (Table 9.1). This is confirmed by the figures from MOFERT, which show that grain exports increased from 1.9 million tonnes in 1983 to 9.1 million tonnes in 1986 (Table 9.3), and the import level decreased from 13.4 to 7.3 million tonnes during the same time period (Table 9.2). Generally speaking, the successive bumper harvests achieved during 1979–84 have fundamentally changed the nation's trade

Table 9.1 China's grain imports and exports, 1983-86

	1983	1984	1985	1986
		– Thousand tonnes –		
Grain exports	1,150	3,190	9,320	9,420
In which:				
Rice	580	1,160	1,010	950
Soybean	350	840	1,140	1,370
Grain imports	13,530	10,410	6,000	7,730
In which:				
Wheat	11,110	9,870	5,410	6,110
Barley	70	50	32	199
Corn	2,110	60	90	588
Beans	40	40	41	340

Note: Figures presented in this Table are from the customs which is different from the foreign trade department.
Sources: Figures for 1983, 1984, *TJNJ 1985*, pp501, 506.
Figures for 1985, 1986, *TJNJ 1987*, pp597, 600.

Table 9.2 The composition of China's grain imports, 1983-86

	1983		1984		1985		1986	
	Volume (100 tonnes)	% of total	Volume (100 tonnes)	% of total	Volume (100 tonnes)	% of total	Volume (100 tonnes)	% of total
Grains	134,351	100.0	106,454	100.0	61,711	100.0	72,824	100.0
In which:								
Wheat	110,191	82.0	100,000	93.9	56,324	91.3	57,535	79.0
Corn	19,863	14.8	465	0.4	800	1.3	6,831	9.4
Rice	1,608	1.2	2,461	2.3	3,132	5.1	3,190	4.4
Barley	1,066	0.8	1,014	1.0	399	0.7	3,904	5.4
Glutinous rice	383	0.3	100	0.1	–	–	200	0.3
Green bean	347	0.3	413	0.4	430	0.7	392	0.5
Others	893	0.6	2,001	1.9	626	0.9	772	1.0

Note: Figures presented are from the foreign trade department and not the customs.
Source: Bureau of Agriculture, the State Planning Commission of China.

Table 9.3 The composition of China's grain exports, 1983–86

	1983		1984		1985		1986	
	Volume (100 tonnes)	% of total	Volume (100 tonnes)	% of total	Volume (100 tonnes)	% of total	Volume (100 tonnes)	% of total
Grains	19,630	100.0	32,270	100.0	88,800	100.0	90,940	100.0
In which:								
Rice	5,660	28.8	11,890	36.8	10,190	11.5	9,560	10.5
Corn	610	3.1	9,110	28.2	59,570	67.1	57,060	62.7
Sorghum	160	0.8	560	1.7	3,030	3.4	4,030	4.4
Others	13,200	67.3	10,710	33.3	16,010	18.0	20,290	22.4

Notes: a. Figures for other grain exports include the entrepôt trade of grains
b. All the figures presented are from the foreign trade department which is different from that of the Customs.

Source: Bureau of Agriculture, the State Planning Commission of China.

position from being a significant net grain importer in the early 1980s to a balanced grain trade with even slight net exports in the mid 1980s.

Second, the pattern of grain imports and exports has also experienced a profound change. The most obvious change in trade pattern has been in the balance of composition of grain exports. Traditionally, rice and soybean have been the two most important products exported, accounting for around 80% of the total (see Table 7.2). However, the proportion of rice and soybean exports declined sharply to 62.7% in 1984 and even lower in 1985 to 23.1% and 24.6% in 1986. The main reason for this dramatic structural change is revealed by Table 9.3 which shows that the rapid increase in grain exports since 1984 was mainly accounted for by the increase in corn exports. These accounted for only 3.1% of total grain exports in 1983, but rose to 28.2% in 1984 and 67.1% in 1985. In terms of grain imports, Table 9.2 shows that wheat remains the most important import product despite the fact that domestic wheat production registered a 34 million tonnes net increase in output during 1978–1984 with an annual growth rate of 8.5%. Wheat import made up 82% of the total grain imports in 1983, it went up to 93.9% in 1984 and 91.3% in 1985. In contrast, imports of corn accounted for 14.8% of trade in 1983, and has subsequently fallen to an insignificant proportion of 0.4% for 1984 and 1.3% for 1985. The 9.4% registered in 1986 is still well below the corresponding figure for 1983.

For many economists, the changes in China's grain trade pattern looks rather odd contradicting their general expectations. Indeed, it is generally believed by economists that given a recovery to the per capita food grain consumption levels of the early 1980s, any further increase in demand for grains would be mainly manifested as increased demand for feed grain enabling dietary improvement. Therefore, the general expectation for changes in China's grain trade pattern was an increase in rice exports, a reduction in the import of wheat, and a sharp increase in the import of feed grain, especially corn. Indeed, studies of China's future grain trade so far published all emphasized a fundamental change in grain import composition in the near future; namely the substitution of feed grains for wheat in China's grain imports. These studies also expected a significant increase in China's rice exports, pointing out the main constraints in the world market (see, for example, World Bank 1985b and Carter and Zhong 1988).

It is useful to try to understand the reason events did not turn out as expected. The recent changes in China's grain trade pattern are determined mainly by two key factors. One is the change in food grain consumption composition. For most Chinese people, dietary improvement during recent years has not only been manifested by the consumption of more meat, eggs, vegetables, and fruits, but also, and equally significant, on the changes in the consumption composition of food grains themselves. During the early days, consumption of fine grain was mainly constrained by its availability. Consequently, coarse grain, corn and sorghum in the north and sweet potato in the south, became the main source for human consumption for most rural populations. For urban citizens, not only was the total amount of food grain supply fixed, but also the proportion of fine grain available. The improvement in availability of rice and wheat and the increase in personal disposable income during recent years have enabled Chinese people to consume more fine grain. Consequently, the proportion of fine grain in food grain consumption has increased considerably, both for urban and rural populations. According to Gong (1984, p 136), the portion of fine grain consumption of the nation has increased to 63.6% of the total food grain consumption in 1981 and the corresponding figure for the urban population was 70%. The upward trend has continued during the last few years and is expected to continue for the foreseeable future.

The other factor dictating the changes in trade pattern is the immobility of the nation's grain products. The combination of the great distances involved in internal shipments and weakness in the social infrastructure make long internal transhipments extremely costly and in some cases unworkable. Compounding these difficulties is the geographical distribution of crops. Fine grain, typically rice, is mainly produced in south China, and coarse grains, mainly corn and sorghum, are predominantly produced in the north, especially in the three northeastern provinces of Liaoning, Jilin and Heilongjiang. Following the recent bumper harvests in grain production, more rice in producing regions was used both as food for direct consumption and as feed to support animal production. This is one of the main reasons that the significant production increase has not resulted in more rice being available for export. By the same token, the increase in coarse grain production in the northeast of China did not increase the availability of coarse grain for feeding animals in the south but rather boosted export to the world market. The sharp increase in corn and sorghum exports during 1984–86 shown in Table 9.3 is the direct result of production growth in the northeast, especially from the corn production region in Jilin Province. It is suggested that these two basic underlying factors will continue to condition the further development of the nation's grain trade pattern. This issue will be discussed further below after considering recent development in the sources and destinations of China's grain imports and exports. The destinations for China's rice exports during 1984–86 are shown in Table 9.4 and the sources for grain imports in general and wheat imports in particular are presented in Tables 9.5 and 9.6 respectively. As can be seen from Table 9.4, there is a wide and diversified range of destinations for China's rice exports during recent years. The marginal reduction in the total amount of rice exported during this period is due primarily to the availability of rice in China rather than any limitations in the world market. It is therefore suggested that, in the foreseeable future, the amount of rice exported from China will most probably continue to be constrained by domestic availability. From the world market point of view, there is still room to manoeuvre. The principal sources of China's wheat imports have been Canada, United States and Australia until 1984 (see Table 9.6). Nevertheless, significant changes have also occurred since 1985. The amount of wheat imported from the United States was drastically reduced in 1985 and went down further in 1986 to the insignificant volume of 60,000 tonnes. France and Argentina have become important suppliers of China's wheat imports since 1985.

9.2 Views on China's Prospective Position in the World Grain Market

In prospecting the future development of the nation's grain trade and her prospective position in the world market, the International Trade Research Institute

Table 9.4 The destinations of China's rice export, 1984-86 (In 100 tonnes)

	1984	1985	1986		1984	1985	1986
Total amount	11,891	10,190	9,565				
Hong Kong	1,754	1,500	1,221	Poland	328	699	503
Macao	124	115	107	Hungary	50	99	90
Singapore	30	27	44	Czechoslovakia	498	495	407
Sri Lanka	–	1,315	919	Bulgaria	–	100	100
Iran	–	–	50	Albania	–	–	47
Turkey	–	100	20	Romania	339	303	305
The PDR of Yemen	400	306	304	GDR	180	202	261
Balin	28	28	49	Yugoslavia	–	–	20
Qatar	9	10	13	France	2,198	458	415
Saudi Arabia	109	92	76	Great Britain	–	–	105
Oman	39	32	37	Sweden	5	12	5
Libya	198	298	403	Switzerland	1,284	1,621	1,139
Costa Rica	–	–	262	Cuba	330	505	1,006
Gabon	30	3	3	Brazil	30	596	–
Ethiopia	–	–	10	Peru	–	–	610
Tanzania	158	–	–	Fiji	–	9	–
Mauritius	392	371	387	Others	3,399	1,469	51

Note: Figures presented are from the foreign trade department and not that of the Customs.
Source: Bureau of Agriculture, the State Planning Commission of China.

Table 9.5 Sources of China's grain imports, 1984-86

	1984		1985		1986	
	Volume (100 tonnes)	Value (1000 US$)	Volume (100 tonnes)	Value (100 US$)	Volume (100 tonnes)	Value (1000 US$)
Total amount	106,454	1,611,720	61,711	857,498	72,824	819,083
Hong Kong	50	959	44	1,115	1,921	21,835
Macao	7	256	–	–	1	53
Korea	306	4,763	784	13,769	194	3,653
Japan	1,755	18,674	558	9,021	495	6,173
Burma	354	6,310	882	14,937	931	13,853
Thailand	2,754	65,862	2,706	54,719	6,978	92,086
Nepal	11	265	13	364	6	86
France	–	–	2,770	30,974	2,944	30,558
Argentina	–	–	6,863	79,837	5,494	52,420
Canada	30,785	459,995	22,902	327,899	24,956	274,329
U.S.A.	45,070	663,972	11,540	155,008	1,165	12,338
Australia	24,410	373,236	12,649	169,854	27,738	311,700
Others	952	17,428	–	–	–	–

Note: Figures presented are from the foreign trade department and not from the Customs.

Table 9.6 Sources of China's wheat imports, 1984-86

	1984		1985		1986	
	Volume (100 tonnes)	Value (1000 US$)	Volume (100 tonnes)	Value (1000 US$)	Volume (100 tonnes)	Value (1000 US$)
Total amount	100,000	1,495,560	56,324	758,990	57,535	643,701
Hong Kong	–	–	–	–	7	58
France	–	–	2,770	30,974	2,944	30,558
Argentina	–	–	6,863	79,837	5,194	49,675
Canada	30,524	456,351	22,901	327,899	21,787	250,633
U.S.A.	45,070	663,972	11,540	155,008	600	7,766
Australia	23,657	362,360	12,249	165,272	27,003	305,011
Others	749	12,877	–	–	–	–

Note: Figures presented are from the foreign trade department.
Source: Bureau of Agriculture, the State Planning Commission of China.

(hereafter ITRI) of the MOFERT, which is the most influential research institute concerned with trade affairs in China, spelled out the general strategy for a rational trade pattern for China as follows. They argued that, from the long-term point of view, the basic strategy is to feed over one billion Chinese people. This, in turn, suggests priority be given to the encouragement of further domestic grain production instead of relying heavily on importation from the world market. This argument was justified by both economic and political considerations. From the economic point of view, it was expected that the world market for grain products will continue to involve great uncertainty, possibly with substantial fluctuations in price. It was argued that on average, the import of one tonne of grain from the world market has historically cost the Chinese government the equivalent expenditures as purchasing two tonnes of grain produced domestically. Therefore, more grain imports from the world market would inevitably require more financial subsidies from the state. Furthermore, since China has set a course for modernization, the priority use for hard currency is to import more new technology and modern production facilities. It is therefore regarded as unjustifiable to spend additional foreign exchange to import large amounts of grain products. This stance is maintained despite the fact that the price gap between domestic purchasing and that of the world market has narrowed considerably. The political element in China's position on international grain trade is also regarded as important. It is argued that the domination of the world market by the United States is a harmful element for the stabilization of the world market. Historically, the American government had not only used grain supply as a political weapon with the USSR, but also, and quite often, as an effective means to put political pressure on developing countries for whom the world market supply is vital for food supplies. These renewed uncertainties have encouraged an attitude of greater self-reliance for grain supply.

In projecting the nation's general grain trade position to the end of this century, the ITRI expected that China will continue to import a certain amount of grains from the world market to balance her domestic production and utilization. During the 1980s, the annual net import level will probably be maintained at 10–15 million tonnes. For the 1990s, provided the domestic production growth rate can be kept at 2.1% per annum and the 1% natural growth rate of population set by the government is not exceeded, the nation will then most likely remain as a net importer but with a reduced level of import. As China was faced not only with a shortage in supply of grain, but also cotton and sugar in the early 1980s, the ITRI argued that, from the point of view of economic rationality, it would be better for the nation to import more grain to support domestic cotton and sugar production. This was based on the fact that price movements have worked in the direction of increasing the costs of imported cotton and sugar. For example, the cost of one tonne of sugar was equivalent to the importation of 1.75 tonnes of grain in the 1960s, however, the ratio exceeded 1:2.25 in the 1970s. In terms of the composition of grain trade, the ITRI called for increase in export of rice and soybeans. It was believed that the main measure to encourage expansion of rice exports is to improve its quality, making it more competitive. For the soybean exports, the target set by the ITRI was net export of one million tonnes for 1990, and two million tonnes by the year 2000. The major destinations targeted were Japan, the USSR and other East European countries (see ITRI 1984, pp 145–60).

Wu Shuo of the Economic Research Institute of China's Ministry of Commerce argued that since agricultural production is subject to substantial yearly fluctuations, it is necessary for the state to set up a "stockpile fund" to store up a significant amount of grain and cotton as the nation's "strategic reserve". At the same time, the state marketing agency has to be able to keep a sufficient amount of "turnover stock", in order to perform its buffer stock rôle properly. Moreover, rural agricultural co-operatives and peasant households have also to be encouraged to stock certain amounts of grain to even out the effects of inevitable fluctuations in yearly production. Wu argued that such reserves would enable the nation to be in a much better position to cope with changing world market conditions and enjoy a greater degree of freedom in pursuing her goal of economic rationality in agricultural trade (Wu 1985, p114).

The discussion above has explained how the combination of regional differences in cropping patterns and the general weakness in social infrastructure in China has played a very important rôle in explaining the recent changes in the pattern and volume of international trade. As it is expected that the improvement of the social infrastructure will continue to be a long and slow process, the future regional cropping patterns will continue to be an important factor in determining the nation's trade pattern in the foreseeable future. It has been projected by the CAAS (1985, pp 13–14) that, provided there are no significant changes in regional cropping patterns, toward the end of the century there will be 6.5–9 million tonnes of surplus grain from the Dongbei (northeast China) and Huabei (north China) regions, two million tonnes from the Huadong (the Middle and Lower Reaches of the Changjiang River) region, while Huanan (Southeast China) and Xibei (Northwest China) regions will face a net deficit of 1.7 and 2.1 million tonnes of grains respectively. From the crop composition point of view, the continuation of the present cropping pattern will create in the north (typically Dongbei and Huabei regions) a shortage of fine grain supply, with a large surplus of coarse grains, especially corn products. In contrast, the south (especially Huanan and Huadong regions) will face serious shortage in coarse grain, whilst producing a surplus in fine grains (rice and wheat). To avoid huge quantities of grain transferring between regions, the CAAS called for an adjustment in regional cropping patterns. That is, to shift the low-yielding rice growing area in the south into the production of corn and other forage crops and to expand rice and wheat production in the north. Their proposal is thus to try and more closely match a particular region's production to its consumption requirement. However, the feasibility of a significant shift in regional cropping patterns and how far it can go in the near future has yet to be made clear. The economic rationality of this notion of regional self-sufficiency also deserves closer study than can be attempted here.

The World Bank has made its projection about China's grain economy towards the end of the century (see World Bank 1985b). In general, it was expected that from the long-term point of view, it is most unlikely that China will be faced with shortages of food grain supply for human consumption. The shortage, it appears, would probably be the coarse grain for livestock production. It was projected that between 30 to 48 million tonnes of surplus rice would be produced by the year 2000. As a result, 30 million tonnes of unmilled rice or 21 million tonnes in milled form would be available for export. Meanwhile, the Bank estimated that domestic consumption of feed grain would exceed production by between 45 and 60 million tonnes and up to 45 million tonnes of imports of feed grain and high protein meals could be required to support growing domestic animal production. The Bank also argued that limitations on port capacity and the distributional infrastructure will remain as the main constraints to the expansion of grain imports and exports. In addition, given the foreign exchange constraint, the capability of the nation to import feed grains will also be determined partially by the extent of the expansion in exports of rice and non-grain agricultural products. It was projected by the Bank that, while the world market price ratio between rice and wheat in the mid 1990s will become more favourable for the nation to export more rice than it stands in the early 1980s, the world market by 1995 can only absorb a maximum of 16 million tonnes of rice. It was also recognized that China's diversity in her endowment of agricultural resources and low cost labour will enable the nation to produce a wide range of high-value and special native products for export. Nevertheless, there is also limitation on world market demand. So, again, a rapid increase in exporting these products from China would depress market prices. In general it has to be concluded that international trade will not provide a unique solution to the structural issues involved in the future development of Chinese agriculture in general and her grain economy in particular. Therefore, the necessity for adjustment in regional cropping patterns was also stressed by the Bank.

Carter and Zhong (1988) also discussed trade issues based upon their projections of China's grain production and consumption. Their conclusions were, first, the nation's demand for grains will exceed its domestic supply in the 1990s, therefore, China is likely to remain as a net grain importer; second, future grain imports of the nation will shift from wheat to feed grains; third, considering the constraints faced by the grain trade, the quantity of feed grain imported will be smaller than that projected by the World Bank.

9.3 Prospects for the Future Trade of Grain Products

In this section the projections of domestic production and consumption which were derived in Chapter Eight are brought together to examine the consequences for grain trade through to the year 2000. The analysis starts with the construction of grain balance tables for the critical years 1990, 1995 and 2000. This is followed by the discussions on the prospects for the net trade position and the scale and composition of grain trading.

9.3.1 The construction of a grain balance table

To simplify the analysis and presentation, only the medium projection scenario for grain production, which is regarded as the most probable outcome, is presented in the balance sheet. The projected grain output is disaggregated into rice, wheat, soybean and other grains. As far as consumption is concerned, domestic consumption of grains (the total utilization) covers direct human consumption (food grain) and indirect consumption of grain in three categories: feed, seed and industrial use. In addition, because wastage of grain between harvest and consumption is a common phenomenon, it is included as a requirement item in indirect consumption. Since there are no official statistics available in terms of real amount of wasted grain of the nation, 0.62% of the total outputs is taken as the estimate based on work by Kueh (1984, Table 3). Adjustment for stock changes is another important dimension of the disposition of grain and it therefore must be taken into account. Because there is so little information on the size of adjustments of grain stocks by peasant households and rural co-operatives, these private stocks are ignored. This is equivalent to making the assumption that these stocks are maintained at a constant level. The supply balance sheet in Table 9.7 therefore includes adjustment in state stocks only. It is estimated that the state grain stock in the mid 1980s has reached the level of around 70–80 million tonnes (milled grain). In this study, while no further adjustment in state stock is assumed for the low scenario, an annual increase of one and two million tonnes of grain is assumed for the medium and high scenarios respectively. Finally, figures for the net import requirements are the balancing results after taking the production, domestic consumption and adjustment in state stock into account. Details can be seen in Table 9.7.

With the three alternative scenarios specified for the consumption of grain, Table 9.7 shows that between 10.8–43.8 million tonnes of grain imports are required to achieve an equilibrium in 1990. For the year 1995, the range is from a net surplus of 3.5 million tonnes to net deficit of 50.4 million tonnes. To the end of the century, the range is further widened. That is, from the net surplus of 17.2 million tonnes to a net deficit of nearly 60 million tonnes. The general indication from these supply balance tables is therefore that the issue of the future balance of China's grain products is quite open; it could result in shifting China from being a net grain importer to a net exporter, or the scale of net imports could grow substantially from their current level. These hypothetical grain balances serve to indicate the range of possibilities for the nation's future grain trade. However, the figures for net import requirements derived in Table 9.7 only reflect the consequences of the alternative combinations of production and consumption under the assumptions specified. Such results cannot automatically be regarded as the possible outcomes of the nation's grain trade. The most important qualification is the fundamental structural change in the composition of China's food consumption, that is the shift from the predominance of direct grain consumption to the improvement of dietary intake. In the event of any possible internal or external disturbances these changes could be reversed or even accelerated, thus affecting the grain balance substantially. Moreover, a set of important factors which influenced grain trade considerably in the past could continue to influence the decisions on future grain trade. These include the nation's balance of payment situation, the domestic grain supply and demand conditions, the world grain market situation and price movements, the internal and external political developments and so on. Rather than subjectively trying to assess

Table 9.7 Alternative projections of China's grain balance for 1990, 1995 and 2000

	1990			1995			2000		
	Low	Medium	High	Low	Medium	High	Low	Medium	High
	(unmilled grain in million tonnes)								
Domestic grain production (Medium)									
Rice	184.0	184.0	184.0	198.6	198.6	198.6	213.3	213.3	213.3
Wheat	81.2	81.2	81.2	88.7	88.7	88.7	96.2	96.2	96.2
Soybean	15.4	15.4	15.4	17.8	17.8	17.8	20.3	20.3	20.3
Others	164.2	164.2	164.2	188.5	188.5	188.5	212.9	212.9	212.9
Total	444.8	444.8	444.8	493.6	493.6	493.6	542.7	542.7	542.7
Direct human consumption	306.3	310.9	314.0	315.2	324.6	331.1	321.2	335.8	346.0
Indirect consumption									
Feed	93.8	107.4	117.1	112.7	134.3	148.8	133.3	163.5	183.6
Seed	29.5	29.5	29.5	29.5	29.5	29.5	29.5	29.5	29.5
Industrial use	23.2	23.2	23.2	29.6	29.6	29.6	38.1	38.1	38.1
Waste	2.8	2.8	2.8	3.1	3.1	3.1	3.4	3.4	3.4
Total	149.3	162.9	172.6	174.9	196.5	211.0	204.3	234.5	254.6
Total utilization	455.6	473.8	486.6	490.1	521.1	542.1	525.5	570.3	600.6
Stock change	0	1.0	2.0	0	1.0	2.0	0	1.0	2.0
Net import requirement	10.8	30.0	43.4	–3.5	28.5	50.4	–17.2	28.6	59.0

Notes: (1) Figures for the domestic grain production are the results of medium projection scenario from Table 8.9.
(2) Total utilization is the sum of direct human consumption and indirect consumption.
(3) The waste of grain concerns the grain lost after harvesting. In this Table, 0.62% of the projected outputs are taken as the estimates of grain wasted, see Kueh (1984, Table 3).
(4) Figures for stock change concerns the assumed annual increase in state grain stocks.
(5) Figures for net import requirement are derived by subtracting total grain outputs from the sums of total utilization and stock added.

the combined effects of these factors and thus choosing the most likely outcome of Table 9.7 the following section considers each issue separately.

9.3.2 Assessment of the future pattern of China's grain trade toward the end of the century

In assessing the future movement of the nation's grain trade within this century, three issues are examined which cover the possible net trade position, the scale of grain trading and its composition.

China's future net trade position. The key question to be addressed is whether China will once again become a net grain exporter. The brief answer to this is that it will be most unlikely. The supporting arguments for this statement follow. First, the essential feature of China is a very large population with limited arable land. Struggling to feed nearly one quarter of the world's population with only 7% of the arable land has always been a formidable task for any regime. This is why China has been a net grain importer throughout most of this century except during the 1950s. During the 1950s, 21 million tonnes (milled) of net grain exports were registered (Table 7.2) based on per capita grain availability of less than 300 kgs (unmilled). This was achieved only by the government's strongest commitment to launch the ambitious industrialization programme under exceptionally favourable internal political conditions. Nevertheless, as was reviewed in detail in Chapter Two, to achieve the 50 million tonnes annual grain procurement and consequently 2 million tonnes of net exports, a fierce political struggle occurred between government and peasant farmers and between central and local governments. It was also one of the main contributors to the failure of the People's Commune movement and to the nationwide famine.

The present socio-political environment in the nation is rather different from that in the 1950s. Current opinion would not tolerate an overemphasis on the development of heavy industry with the neglect of people's daily living standards and the overcentralized control system of the 1950s is nowadays unattainable. Instead, balanced development with a more decentralized management system and the limited introduction of market mechanisms are now pursued. The efforts to improve people's daily living standards is regarded as one of the key elements both to motivate individual initiative in development and to encourage the people's enthusiastic support for the cause of restructuring. Consequently, as described by Surls (1983, p134), "The Chinese government now has a stronger commitment to maintaining and raising standards of living than was the case in the past. Because of this, the government now seems less likely to pressurise rural areas for grain supplies in years of poor harvests. It is also less likely to permit procurement shortfalls to be completely absorbed by lower consumption in the urban/industrial sector or reduced resales to rural areas" (quoted from Kueh 1984, p924). Thus with the abolition of the compulsory purchasing system, even if the government wanted to pursue the old line it would be unworkable in practice. Therefore, the days of forced extraction of grains for export have gone.

Second, following the discussion of the previous chapter, the task for the government to achieve even the medium scenario for production is formidable. It is by no means easy both to maintain the grain growing area at the current level and to induce massive productive investment in the agricultural sector to enable a con tinuation of the current trend of technological improvement. Meanwhile, and more importantly, bearing in mind the dilemma faced in manipulating price policy for future grain production as discussed in section 8.4.1.2, all these imply that a further intensification of domestic production to a level beyond the medium scenario is technically and economically irrational.

Third, it is expected that within this century, population growth will almost certainly again exceed the 1% annual rate set by the government. It is also foreseeable that the growth of GNP and the increase in personal disposable income will go beyond the pace specified by the government (see section 8.3). In addition, the costly state subsidies which make grain cheap and encourage consumption will not be easily ended in the foreseeable future because of its high political sensitivity. Consequently, the current trend of a tremendous pressure of demand for grains, both for food and feed, as well as the situation of domestic production exceeded by consumption will continue for the foreseeable future. Under these circumstances, net imports of grain are absolutely necessary to fill the gap.

In summary, given the social, political and economic issues reviewed, it is believed that China is extremely unlikely to become a systematic net grain exporter once again this century. Instead, the current position of net grain importation will continue for the remainder of the century.

The scale of net grain trading With the continuation of the "open door" policy and continuing efforts in encouraging the further expansion of foreign trade in the future, there will be no obstacle to trade in terms of general socio-political consideration as there was during the Cultural Revolution. Nevertheless, there are three particular factors which will play an important rôle in limiting the growth of net grain imports. First, it is believed that economic rationality will continue to be an important motivation in the nation's manipulation of grain imports and exports. Therefore, China's ability to export rice and other non-grain agricultural products will be vital in determining her capacity to import more grains. If the export of rice in exchange for wheat and corn is to be more attractive as suggested by the World Bank, the expansion of rice will be itself constrained by resource availability as discussed in section 9.1 above. The scope for expansion of processed foods and other agricultural products in similarly limited. The constraints again are not the availability of export markets, but the growing competition of the domestic market shown by increased demand pressure and rising prices. Consequently, the balance of payments would become a critical constraint should a massive increase in net grain imports be required.

Second, another obstacle preventing a considerable increase in grain imports is the limitations on port storage facilities and the distribution system. Although these are currently under improvement, the weakness in infrastructure represents a substantial bottleneck which would inhibit a significant increase in grain imports if this was attempted in the foreseeable future.

Third, most importantly, a huge volume of net grain imports could cause political problems. Because grain is such an important product for the nation both in terms of economic development and on the political battlefield, an obvious failure in boosting domestic production to catch up with consumption requirements would be regarded as evidence of the failure of current socio-economic restructuring. The sensitivity of this issue is clear from the rigorous political debates amongst the top leadership when the issue of the stagnation of grain production is raised. Therefore, in addition to the possible constraints on balance of payments and the weakness in infrastructure as mentioned above, a dramatic increase in net grain imports would be regarded as unacceptable for the administrative system politically. This is far more important than the other factors. It is concluded therefore that the maximum level of net grain import within this decade will be at around 20 million tonnes, and for the 1990s that of 30 million tonnes.

The composition of grain imports During the last three decades, grain imports of the nation were dominated by wheat for the purpose of direct human consumption (see section 7.2). Given that the per capita grain availability has been improved substantially since the end of the 1970s, most researchers have suggested that future grain imports in China will be dominated by the demand for indirect grain consumption, that is import of corn as feed grain would be the predominant component instead of wheat (see for example Kueh 1984, p925; World Bank 1985b, pp35–105; Carter and Zhong 1988). It is suggested here that whilst there will indeed continue to be structural change in the composition of grain imports, wheat will remain a most important component of grain imports for the foreseeable future. This conclusion is justified by three arguments. First, because of the higher population growth rate expected, larger volumes of food grain consumption are projected than in other studies. Second, one of the major characteristics of the Chinese people's dietary improvement during recent years is the increase in proportion of fine grain (rice and wheat) in food grain consumption. It is expected that this trend will continue in the future. Third, historically, imported grains have always been distributed mainly in the eastern coastal areas, and one of the main reasons for this is the weakness in the transportation and distribution system which results in greater costs for wider distribution. Since there can be no realistic expectation of drastic improvement in infrastructure in the near future, fine grain, especially rice, will continue to be used to feed animals in vast rural areas in central and south China, and imported grain will continue to be distributed as food grain in the coastal areas, especially for the three metropolitan cities. In summary, the proportions of wheat and corn in the future grain imports will not only be critically dependent upon the composition of direct and indirect consumptions of grains, but also, and equally important, on the improvement in the grain distribution system which is the key to a more rational use of grains.

9.3.3 Summary

Projections of the future development of China's grain economy have been pursued in the last two chapters, the main findings are summarized below.

First, the general socio-economic environment for future rural development has been set by the second stage rural restructuring. The implementation of the new system will certainly stimulate peasant farmers' enthusiasm for the development of commodity production and continue the current unprecedented flourishing of the rural economy. However for grain production, because of the constraints on arable land and the currently unattractive pricing policy, government intervention will be necessary for a sustained growth of grain production toward the end of the century. Key policy measures required to protect grain production are: (1) stabilizing the growing area at current level both by tightening control on the shift of arable land into other development uses, and by reclaiming new land for crop cultivation; (2) continuing the trend of technological improvement by means of inducing massive productive investment in the grain sector to maintain high growth rate in modern inputs and further improvement in scientific research and extension services; and (3) most importantly, getting the price right to make grain production more attractive for producers. In view of the particular importance of grain products and the government's determination further to boost production, it is believed that the medium projection scenario presented will be the most probable outcome within this century.

Second, it is projected that food and feed will be the two most important components in total grain consumption in the remainder of this century. Mainly due to higher population growth rate than the government target, the current trend of high demand pressure on grain products will continue for many years to come. Therefore it is almost certain that the domestic consumption requirements will continue to exceed the domestic production for the remaining years of the century. To ensure the imbalance does not get out of control, critical policy measures likely to be taken by the government include: (1) continuation of the current efforts of strict control in population growth; (2) regulation of the consumption composition for consumer goods to alleviate the market pressure on foods by means of government intervention in market supply-demand conditions and manipulating in pricing policies; (3) continuation of the current practice of outward looking and encouragement a further expansion in the exports of rice and other agricultural products, in order to mobilize more foreign currency for the imports of grain products.

Third, although the possibility of China becoming a net grain exporter is regarded as most unlikely in the foreseeable future, a massive increase in grain imports is also considered as impossible for various reasons especially because of its high political sensitivity. Therefore, 20 million tonnes of net imports are believed to be the maximum level within the 1980s and 30 million tonnes for that of the 1990s. Meanwhile, it is believed that factors which influenced the historical development of grain trade greatly will continue to play important parts in the future. In addition, it is expected that while corn imports are going to become more important, wheat will remain as a major product in the nation's grain imports in the foreseeable future. Finally, the general conclusion is thus, in regulating the future development of the grain economy, the task for the government will be formidable although attainable.

9.4 Interaction Between China and the World Grain Market

9.4.1 Prospects of the future world market

During the last few years, a number of attempts have been made to project the world market for agricultural produce including grain products up to the end of the century. Before considering China's possible impact on the world grain market, it is necessary first to review briefly studies conducted on the future world market by such influential institutions as the United States Department of Agriculture (USDA), the World Bank and the Food and Agriculture Organisation of the United Nations. These are each considered in turn.

USDA studies A world agricultural trade model (SWOPSIM framework) has been used by the USDA to examine the outlook for agriculture to the year 2000 (see Roningen, Dixit and Seeley, 1988). the model used is a static, non-spatial price equilibrium simulation model. The model specifies 11 regions (seven for developed countries, three for developing countries and one for centrally planned economies) and 22 major agricultural commodities. The principal economic relationships are supply, demand, trade and price equations for each commodity and each region. Data used came from USDA, US Bureau of Census, the IMF, World Bank and FAO.

The base run of the modelling exercises assumed that there would be a continuation of past growth patterns and present policies, i.e. a growth rate of 2.5% per annum for developed countries and higher rate for developing countries; the population growth rate for the developing world was higher than that for developed nations; and the per capita income growth for developed countries would continue to be higher than developing countries. Results from the base run showed that the real aggregate agricultural prices will decline 4% from 1984/85 to the year 2000. Within this, prices for ruminant meat, dairy products and rice will go up, and prices for grains will fall. Global agricultural production was projected to increase at 2% per annum over the period of 1984/85–2000. Per capita food consumption for developed nations will maintain its current level but will rise by 2% per annum in developing countries and centrally planned economies. Moreover, it was projected that the growth of supplies in developing countries would be unable to keep pace with expansion in demand, therefore, their self-sufficiency level for agricultural products will be lower in the year 2000, necessitating increased imports. In contrast, developed regions will improve their balance of trade with a higher degree of self-sufficiency for most products, particularly grains.

One of the main policy simulations carried out in the USDA analysis was an investigation of the effects of a movement towards freer trade. In this, the developed nations eliminate all subsidies which distort agricultural trade as was proposed by the United States for the Uruguay Round of GATT. Results from this simulation showed that the aggregate agricultural price in the year 2000 was 8% higher than that from the base run. Within this overall price rise, prices for sugar, dairy products, ruminant meats, and rice go up, and the rate of decline for grain prices slows down. The results of these changes are that the self-sufficiency ratio for developing countries and centrally planned economies improved as they reduced their imports of agricultural produce. The developed nations experienced a decline in agricultural exports and an increase in imports. The model indicated that this policy change would have little impact on the global agricultural production level although a lower share for developed countries was demonstrated.

The other main policy simulation concerned changes in economic growth. The basic assumptions include higher and lower global economic growth patterns. With the higher growth assumption, the annual growth rates assumed was 1% higher for developing countries and centrally planned economies and 0.5% higher for the developed countries (all compared with the base run). This is similar to the growth pattern registered in early 1970s. Results showed that the global demand for agricultural products would be 5% higher in the year 2000 due to the higher economic growth assumed. A large share of the additional demand comes from third world countries. Meanwhile, price for agricultural products in general would be 12% higher, within which, the increase for dairy products is the largest and that for grain the smallest. In addition, import requirements in developing countries would increase, resulting in increased exports from developed nations.

The alternative lower growth analysis assumed a similar pattern of economic growth to that in early 1980s. That is, one per cent lower than the trend taken in base run. Results were with the slower economic growth assumption that the rate of increase of global demand for agricultural products would be lower than that derived from the base run (which was 2% per annum). It was also shown that the decline in aggregate agricultural prices in the year 2000 would be more than double that projected in the base run. Therefore, prices would trend downward even for animal products. Generally speaking, a 1% swing in world economic growth would be significant enough to make a difference in agricultural prices trending upwards or downwards.

The World Bank studies Akiyama and Mitchell (1988) of the World Bank used a global econometric model to project the world grains and soybeans markets to the year 2000. The study identified 24 countries and regions to cover the global markets for wheat, rice, coarse grain and soybeans. Production, consumption, net trade and stocks by major regions and world prices were the major components of the model. The model was econometrically estimated with time-series data covering the period of 1960–84. It was then used to forecast the future development to the year 2000. Three simulation exercises were conducted: a base run (trend continuation), high demand (high GDP and high population growth) and low demand (low GDP and low population growth).

The result of the base run was that real prices for all the products analysed will decline during 1990–2000. There would be 23% net reduction for wheat, 16% for maize, 13% for rice and 32% for soybeans (all in 1987 prices). During 1985–2000, wheat production of developed countries will increase 2.4% per annum and coarse grain by 1.7% per annum. A large production increase for developing countries was expected but the rate of growth is still lower than that registered between 1970 and 1985. For the centrally planned economies, while an increase in coarse grain production was projected, wheat production will remain more or less constant. In terms of consumption, a continuation of the past trend was projected for the developed nations, e.g. an annual increase of 1.4% for wheat, 1.8% for coarse grains and 3.3% for soybeans. The consumptions of developing countries were projected as, 3.5% annual increase for wheat, 4.2% for soybeans and 2.4% for coarse grain and rice. The annual consumption increases for centrally planned economy were projected at the range of 2.3–3.3% for coarse grains, rice and soybeans and 0.3% for wheat. Projections of the volume of trade showed that exports of wheat, coarse grain and rice from the developed countries will increase substantially between 1990 and 2000. The magnitudes were an increase of 134 million tonnes for wheat exports, 91 million tonnes of coarse grains and 5 million tonnes of rice in the year 2000. At the same time, soybean imports by the developed countries were projected to fall from 5.3 million tonnes in 1990 to 3.2 million tonnes in the year 2000. From the developing countries' point of view, the net import requirement for wheat would be 72 million tonnes for 1990 and 115 million tonnes for the year 2000, with an annual growth rate of 5% during 1985–2000. The next import requirement for coarse grains was projected at 30 million tonnes for 1990 and 58 million tonnes for the year 2000. The figure for rice was 1.3 million tonnes in 1990 and 5.0 million tonnes in 2000. For the centrally planned economies, the net imports of wheat and coarse grains were projected as 14 and 16 million tonnes in 1990 and 19 and 33 million tonnes in the year 2000 respectively.

Results from these two simulations showed that, generally speaking, production response to the changes in global demand is greater for developed countries than that for developing countries and centrally planned economies, mainly due to the availability of idle land. With the high demand simulations, results showed that wheat consumption in developed countries would fall, the consumption for coarse grains and soybeans would increase and rice remain basically unchanged, all compared with results from the base run. For developing countries, the high demand simulation suggested a significant increase in requirements for wheat, soybeans and coarse grains. The implication for the centrally planned economies was an overall increased demand for grain products. That is, the annual growth rate of demand would increase from 0.3% in base run to 1% for wheat, from 2.3% to 2.9% for coarse grains and from 3.2% to 4.4% for soybeans. In contrast, results from the low demand simulation showed either unchanged demand compared to the base run projection or a fall in demand (for details, see Akiyama and Mitchell, 1988, Tables 5, 6 and 7).

FAO Studies In its publication entitled *Agriculture: Towards 2000* the FAO updated its study on global agricultural development toward the end of the century with new evidence up to the mid 1980s (see FAO 1987). The study involved detailed investigations concerning production, consumption and international trade. Individual countries are grouped into developing countries, European centrally planned economies and the developed market economies. The main conclusions for the grain sector from this study are outlined below.

It was projected that total demand for cereals (including direct and indirect consumption) for 94 developing countries (including China) will reach the level of 1,250 million tonnes by the year 2000, a net increase of 430 million tonnes from the level registered in 1983/85. Cereals production, which is 762 million tonnes in 1983/85, will rise to 1,154 million tonnes in the year 2000. Since the 2.7% annual increase in demand projected is higher than the 2.6% growth rate in production, it was demonstrated that the cereals deficits, that is, the gap between domestic production and consumption for these countries as a whole, will continue to widen. A net deficit of 95 million tonnes is projected for the year 2000 compared to 61 million tonnes recorded in 1983/85. As China and India together account for 55% of grain production and 52% of the consumption of the developing countries, a proper perspective of the development of the grain economy in these two countries was regarded as vital to draw a general picture for developing countries as a whole. The above projection assumed that both China and India will maintain their position of near self-sufficiency in grains.

It was estimated that the total demand for grains in the Eastern European centrally planned economies will be at around 370 million tonnes by the year 2000, an increase of 65 million tonnes from the level in 1983/85 and implying an annual growth rate of 1.2 % during the period 1983/85 to 2000. The total production of this group was estimated as 330–340 million tonnes for the year 2000, which implies an annual growth rate of 1.3–1.5% for the period 1983/85 to 2000. The estimated net import requirement towards the end of the century ranged between 30 and 40 million tonnes which is below the 1983/85 average level of 42 million tonnes.

The aggregate domestic demand for cereals of the developed market economies (excluding stock changes) was projected at the level of 520 million tonnes by the year 2000, a net increase of 65 million tonnes from the average level in 1983/85. The implied annual growth rate during 1983/85–2000 is 0.8%. Production of cereals, even with a rather conservative view, was projected at the level of 796 million tonnes by the year 2000, which implied an annual growth rate of 2% during 1983/85–2000. After the deduction of 522 million tonnes of domestic consumption and 147 million tonnes of net export projected in this study, it was shown that production would still exceed total consumption by 120–130 million tonnes in the year 2000. The general indication of this study was that without a fundamental policy change or systematic negative effects of, for example, the greenhouse effect, the production surplus of the developed market economies will steadily worsen.

In summary, all the studies reviewed indicated that, if there would be no significant change in world economic growth and policies within this century, the market prices for grain products will continue to decline, the gap between grain production and consumption for developing countries will widen. This gap will be filled primarily by increased exports from the developed countries. Meanwhile, future international grain markets will be critically influenced by the possible changes in global economic growth pattern as well as the policy for agricultural trade amongst the significant parties. Higher economic growth will generally result in greater demand for grain and agricultural products, which will push the market prices up. Most of the additional demand will come from the third world countries as well as the centrally planned economies. Increased import requirements from those two groups can only be filled by increased exports from the developed nations. In contrast, lower economic growth would result in lower global demand. Consequently, the current trend of decline in real prices for grains would be further accelerated. Furthermore, should there be fundamental policy shift from current high protectionism pursued by the developed countries, market prices for grain would be significantly increased. Such a new market situation would encourage both developing countries and centrally planned economies to boost their domestic production, thereby improving their self-sufficiency rates and reducing their net imports. Exports from the developed countries, in turn, would be considerably reduced and grain imports increased.

9.4.2 Assessment of China's possible impacts on the international grain market

The global models reviewed above did not especially differentiate China. Their conclusions can therefore only be treated as most general indicators. In this final section the projection of Chinese production, consumption and net trade are considered in the context of likely world market developments.

To assess China's possible impacts on the international grain market, the ideal way would be to run an existing global model using the projected results from this study as the main inputs, to examine their possible impacts on the market supply, demand, trade and prices. It did not prove possible to do this. Nevertheless it is useful to draw together in a qualitative way the results of the projections for China and the studies of world market developments.

It is suggested that, considering all the evidence above, there will be no substantial impacts on the world grain market solely due to the foreseeable changes in China's domestic grain supply and demand. This assertion is justified as follows. First, there is a consensus of analysts of the international grain market that, given the continuation of the current international market developments and trade policies, the widening gap between developing countries' domestic production and consumption of grains can relatively easily be covered by increased exports from developed countries. World market prices for grains will therefore continue to fall and developed countries' grain surpluses, after meeting domestic and export demand, will continue and possibly even worsen. Second, it has been argued in this book that China will most probably remain as a net grain importer for the rest of this century. The grain deficit for the year 2000 was calculated to range up to 59 million tonnes (see Table 9.7). Arguments were constructed taking all important factors into account, to demonstrate that China's net import of grains in the 1990s were unlikely to exceed 30 million tonnes. Both of these figures are significantly lower than the projected net surplus of more than 100 million tonnes by the developed countries by the end of the century.

Furthermore, in contrast to other studies of China's grain trade, it was concluded that while imports of feed grains (mainly corn) could increase in importance, wheat will remain as the most important grain import for the foreseeable future. China is most unlikely to be able to mobilize sufficient rice to export in exchange for the amount of corn suggested by the World Bank. The expected diversification of the pattern of grain trade also makes it less likely that changes in China can have significant impacts on individual product supply and demand conditions in international markets.

More general arguments to support the above assertion concern the security of grain supplies in China and her way of dealing with threats to that security. There is little doubt that current grain consumption levels do provide a more secure base for Chinese agricultural planners to work from than previously. This gives greater flexibility to cope with internal or external market shocks than previously. In the event of a threat to supplies the first response is to internalize the problem in some way or another. This could be manifest as chaos in internal distribution, or a balance of payments crisis or fierce social and political unrest. All of these "coping mechanisms" would be triggered before a profound impact was made on international grain markets. However, it is acknowledged that the analyses reported here are capable only of reflecting long-term trends. Aggregate analysis of this kind is not a useful approach for dealing with year to year uncertainties or shocks. Clearly there can be considerable variability around the "trends" estimated. For some grains, like rice, where world markets are small in relation to global production, such variation can have large effects on world market prices and thus the trade of individual countries. A book of this nature is not the appropriate vehicle for analysing such short-term issues, but it is important to recognize that significant variation around the estimated trends can occur. The general conclusion of this study is thus, within this century, expected changes in China's grain sector alone would not make significant impacts on global grain markets in terms of supply and demand balance and price movements.

It cannot be denied that a number of unexpected fundamental changes inside and outside China could bring significant changes in the world market situation, perhaps even resulting in a global food crisis. First, it is still possible, although most unlikely, that the developed countries might fundamentally change their current highly protectionist policies towards agriculture. There has certainly been a good deal of discussion internationally in the context of the Uraguay Round of the negotiations in the General Agreement on Tariffs and Trade. Such a policy change alone could make considerable differences in global food supply and demand conditions and thus prices. Second, the food situations in other major developing countries and for centrally planned economies could take an unexpected turn for the worse. This could result either from man-made policy disasters or natural catastrophe (drought, flood). Unexpected food crisis from either of these two groups could, no doubt, create serious problems for the world market. Third, the situation in China might run beyond the Chinese government's control. This could take the form of a loss of control of the nation's population growth, or the encouragement of an increase in grain production. A repeat of the social turmoil of the Great Leap Forward or the Cultural Revolution in China which seemed inconceivable until June 1989 would result in a serious food crisis in the world's most populous nation should it occur. This would not only create a serious threat to the ruling regime, but also could have a profound impact on global food markets. Although these occurrences are regarded as most unlikely they cannot be ruled out entirely. Assessing their likelihood and possible consequences is well beyond the scope of this book.

Chapter Ten

The Challenge Remains

The primary focus of this book has been an attempt to understand the economic forces which have shaped the grain economy of China in the tumultuous period since the early 1950s. The substance of the book has been detailed analysis of aggregate statistics of production and consumption of grains in the context of the continually changing general economic circumstances and the sometimes wildly fluctuating policies towards agriculture. Having achieved, since 1978, a period of relative stability rewarded with great success in terms of improved agricultural productivity, it seemed scarcely possible that the gyrations of the past with their disastrous results could be repeated.

At the time of writing it is still too soon after the tragic events of June 1989 in Beijing to assess their long-term significance for the conduct of economic and agricultural policy. This chapter therefore tries only to summarize the unresolved issues of the reform programme at the mid-point of 1989.

China is clearly at a critical stage in carrying out her reform programme. There still exist several important issues to be resolved in the development of the grain economy. Some of these are a direct result of the implementation of the second stage rural restructuring programme, some are more related to the general situation of the nation's socio-economic development. A broad picture of the issues concerning the grain sector is summarized as follows.

First, since the mid 1980s, following management reforms undertaken in the urban economy the costs of industrial inputs like chemical fertilizer, plastic sheeting (used for crop protection), diesel and pesticides have been steadily increasing. These significantly and continuously push up the production costs of grain as well as other major agricultural products. Meanwhile, the encouragement of a reasonably balanced development of the rural economy has opened up more and more opportunities for rural labour and other economic resources. This implies that the opportunity cost for the whole range of rural economic resources in grain production has increased. At the same time, the implementation of contracted grain purchasing has reduced marginal profits for grain production, at least as far as grain which is officially purchased is concerned. Therefore, grain growing has become a less and less attractive economic activity for agricultural producers, consequently peasant farmer's enthusiasm for grain production has considerably diminished. This has been manifested not only in a sharp reduction in grain growing area, but also reduced producers' investment in maintaining and improving production facilities (irrigation facilities, land reclamation and improvement). These effects have been compounded by less favourable weather conditions for crop production especially during 1988. Hence, grain production has stagnated during the late 1980s with total output lower than the record level of 407 million tonnes achieved in 1984. In 1988, the nation's total grain output was 394 million tonnes, down by 2.2% from the previous year. Consequently, China was faced with a net grain shortage of 20 million tonnes. The consequence has been that an estimated 40 million rural people do not have enough to eat.

Second, as mentioned above, the annual amount of grain procurement under contracted purchasing has reduced from 80 million tonnes in 1985 to around 50 million tonnes in 1987. The twin objectives of this move were to provide better incentives for producers and to cut down the budgeting subsidy on grain consumption. In 1988, the direct control of grain supply and demand in Fujian and Guangdong Provinces has been replaced by the operation of the market mechanism. In the long term, these steps can be seen as movements towards basing the whole nation's grain distribution system on the operation of a market system with government intervention relegated to ensuring security of supplies through adjustments in buffer stocks and manipulation of grain imports and exports. Such a fundamental transition would, no doubt, have a large influence on domestic grain production and consumption. However, it is far from clear that there is a determination to make such a transition, nor is it clear how long it would take.

Third, in 1988, there were more than 10 million tonnes of net grain imports. It is expected by the United States Department of Agriculture that China will become the world's largest single wheat importer and the largest single market for U.S. wheat. However, such imports necessitate hard currency earnings via exports. Unfortunately, the combination of high inflation, a strong domestic market and a fixed exchange rate have made it extremely difficult for Chinese traders to mobilize more resources for export. Currently, increased exports by trade corporations, generally speaking, create financial losses which are no longer completely covered by the government. Even for traditional export items such as rice, it has become very difficult to mobilize domestic resources. This situation is worsening. In addition, 1990 is the year China is due to start repaying her massive foreign borrowing. Therefore, increased exports to earn more foreign exchange are required by the government. Even before the political disruptions in June 1989, official spokesmen were admitting that the performance of foreign trade was not sufficient. The medium- and long-term future for China's foreign trade situation is now extremely uncertain.

The steps taken by the government during 1989 have been to raise the official purchasing prices for grain and cotton as well as to increase government investment in agriculture. The aim is to rectify the alarming recent tendency for the production of grain and other major agricultural produce to fall. Provided there were no big natural disasters during the year, grain production in 1989 could be better than the previous year. Nevertheless, from the long-term development point of view, the most important issue remains, getting price signals right to ensure a sustained development of Chinese agriculture and the whole national economy. How best to achieve this was the subject of two sharply differing views. One is the idea of a five year transition in the economic management system reflecting the views of the ex-Party General Secretary, Mr Zhao ZiYang. He argued that price and wage reform was essential to deepen the economic reform programme, and that such was the magnitude of the present distortion that getting price signals correct would have to be phased over a two year period starting at the end of the 1980s. In this case it would take until the mid 1990s to establish a market mechanism for grains. The intended economic outcome of this process were: faster GNP and population growth; a relative rise in grain product prices and, subsequently, in grain yields; a continued fall in grain growing area; and widened scope for grain traded, especially in the coastal areas.

The other view supported by the Premier Li Peng and others was that the main task was to control inflation and to stabilize the economy. Therefore, the transition period for economic management would take ten years rather than five. In this plan with the much longer transition period, the pace of price and wage reforms would clearly be much slower. It would still be possible that a market mechanism for the grain sector would be established but it would not be achieved until around the end of the century. The possible outcomes compared to the previous scenario would be: the GNP and per capita income growth would be lower; population growth rate reduced; there would be a smaller reduction in grain growing area, a less significant improvement in price and therefore a smaller increase in yield; the scope for grain trade would become smaller.

The current sign is that the transition period of economic management will probably take ten years or even longer. The serious social corruption has created social unrest and once again taught the Chinese that economic reform, particularly the development of market orientated economy, cannot be successful without political democratization. Therefore, the drive for four modernizations cannot be achieved without systematic socio-economic and political reform. Just as the necessity of the development of a commodity economy has been widely accepted throughout many centrally planned economies, it has also been perceived by the Soviet Union and Eastern European countries that some democratization of the political system is a co-requisite. This lesson has still been resisted by the Chinese leadership.

Unfortunately the debate between these viewpoints has not been peaceably resolved. The very tensions of the economic reform process itself were manifest as serious social corruption and bureaucratic profiteering. This lead to a rapid and worsening rate of price inflation causing the government to lose confidence in the process of radical price and wage reform. This pause suggested that the more conservative view was gaining the upper hand. Matters came to a head in June 1989 as many people proclaimed the view that the economic liberalization could only succeed if it were accompanied by political democratization. The pace of such reforms demanded was unacceptable to the Chinese government and the world witnessed the methods used to enforce this view.

Whatever the outcome of this struggle during the 1990s the greatest challenge of all, feeding 1.1 billion Chinese, remains. The purpose of this book has been to demonstrate that with appropriate economic management this task is not impossible.

Appendices

Appendix One

Statistical Tables

A1.1 Grain Output in China, 1952-84 (in 10,000 tonnes)

Year	Total grain output	In which: Rice	Wheat	Soybean	Other grain products
1952	16,392	6,843	1,813	952	6,784
1953	16,683	7,127	1,828	993	6,735
1954	16,952	7,085	2,334	908	6,725
1955	18,394	7,803	2,297	912	7,382
1956	19,275	8,248	2,480	1,024	7,523
1957	19,505	8,678	2,364	1,005	7,458
1958	20,000	8,085	2,259	867	8,789
1959	17,000	6,937	2,218	876	6,969
1960	14,350	5,973	2,217	639	5,521
1961	14,750	5,364	1,425	621	7,340
1962	16,000	6,299	1,667	651	7,383
1963	17,000	7,377	1,848	691	7,084
1964	18,750	8,300	2,084	787	7,579
1965	19,453	8,772	2,522	614	7,545
1966	21,400	9,539	2,528	827	8,506
1967	21,782	9,369	2,849	827	8,737
1968	20,906	9,453	2,746	804	7,903
1969	21,097	9,507	2,729	763	8,098
1970	23,996	10,999	2,919	871	9,207
1971	25,014	11,521	3,258	861	9,374
1972	24,048	11,336	3,599	645	8,468
1973	26,494	12,174	3,523	837	9,960
1974	27,527	12,391	4,087	747	10,302
1975	28,452	12,556	4,531	724	10,641
1976	28,631	12,581	5,039	664	10,347
1977	28,273	12,857	4,108	726	10,582
1978	30,477	13,693	5,384	757	10,643
1979	33,212	14,375	6,273	746	11,818
1980	32,056	13,991	5,521	794	11,750
1981	32,502	14,396	5,964	933	11,209
1982	35,343	16,124	6,843	903	11,474
1983	38,728	16,887	8,139	976	12,726
1984	40,731	17,826	8,782	970	13,153

Notes: According to the SSB (1987, p218), the grain output statistics cover the grain produce of the whole society. It includes not only the individual crops like rice, wheat, corn, sorghum, millet and other coarse grains, but also that of tubers and soybean. Figures for tuber output involve sweet potato and potato but not that of Chinese yam and cassava. The conversion ratio from tuber to grain equivalent was 4:1 before 1963 and 5:1 since 1964. In terms of soybean, the conversion ratio was 1:1 to the grain equivalent. All the figures in the table are in the form of unmilled grain. Figures for other grain products are derived by deducting the outputs of rice, wheat and soybean from the correspondent total grain output.

Sources: Figures up to 1982 are from *TJNJ 1983*, p158

Figures for 1983 and 1984 are from *TJNJ 1985*, p255

A1.2 Grain Growing Area in China, 1952-84 in 10,000 Mu (15Mu=1 hectare)

Year	Grain growing area	In which:				Growing area suffering from natural disasters
		Rice	Wheat	Soybean	Other grain crops	
1952	185,968	42,573	37,170	17,519	88,706	6,645
1953	189,955	42,482	38,454	18,543	90,566	10,620
1954	193,492	43,083	40,451	18,981	90,977	18,885
1955	194,759	43,760	40,109	17,163	93,727	11,805
1956	204,509	49,968	40,908	18,070	95,563	22,845
1957	200,450	48,362	41,313	19,122	91,653	22,470
1958	191,420	47,873	38,663	14,326	90,558	11,750
1959	174,034	43,551	35,362	14,795	80,326	20,595
1960	183,644	44,411	40,941	14,022	284,270	37,470
1961	182,165	39,414	38,358	14,936	89,457	43,245
1962	182,431	40,402	36,113	14,256	91,660	25,005
1963	181,112	41,573	35,657	14,450	89,432	30,030
1964	183,115	44,410	38,112	15,013	85,620	18,960
1965	179,441	44,737	37,064	12,889	84,751	16,830
1966	181,482	45,793	35,878	12,638	87,173	14,640
1967	178,845	45,654	37,949	12,754	82,488	–
1968	174,236	44,841	36,987	12,544	79,864	–
1969	176,406	45,548	37,743	12,493	80,522	–
1970	178,901	48,537	38,187	11,978	80,199	4,950
1971	181,269	52,377	38,459	11,687	78,746	11,175
1972	181,814	52,714	39,453	11,375	78,272	25,770
1973	181,734	52,635	39,658	11,112	78,329	11,430
1974	181,464	53,268	40,592	10,892	76,712	9,795
1975	181,593	53,593	41,491	10,498	76,011	15,360
1976	181,115	54,326	42,626	10,036	74,127	17,160
1977	180,600	53,289	42,098	10,267	74,946	22,740
1978	180,881	51,631	43,774	10,716	74,760	32,700
1979	178,894	50,809	44,035	10,870	73,180	22,680
1980	175,851	50,818	43,842	10,840	70,351	33,480
1981	172,437	49,942	42,460	12,035	68,000	28,110
1982	170,094	49,584	41,912	12,622	65,976	23,985
1983	171,071	49,705	43,575	11,351	66,440	24,300
1984	169,326	49,768	44,365	10,929	64,264	22,890

Notes: Figures for other grain crops are derived by deducting the growing areas of rice, wheat and soybean from the correspondent total grain growing area.
Figures for the growing area suffering from natural disasters are the total cropping area suffering from drought, flood,low temperature injury and frost damage and so on with 30% or more net reduction in yield.
" - " not available

Sources: Figures up to 1982 are from *TJNJ 1983*, p154, 155
Figures for 1983 and 1984 are from *TJNJ 1985*, p252, 253

A1.3 Grain and Cotton Purchasing and Fertilizer Selling Prices 1952–84

Year	(a) Grain purchasing price (yuan/ tonnes)	In which (b) Rice (yuan/ 50kg)	Wheat (yuan/ 50kg)	Soybean (yuan/ 50kg)	(c) Cotton purchasing price (yuan/ 100kg)	(d) Fertilizer selling price (yuan/ tonnes)	(e) Gross retail price index (1952 =100)
1952	138.4	5.67	8.15	6.57	183.0	370	100.0
1953	157.2	6.05	9.41	8.01	173.0	364	103.4
1954	157.0	6.03	9.02	7.89	175.8	332	105.8
1955	157.0	6.03	8.94	7.84	179.6	341	106.9
1956	160.2	6.13	8.90	8.15	179.6	341	106.9
1957	162.0	6.18	8.93	8.20	179.6	320	108.5
1958	168.0	6.35	8.98	8.82	178.0	300	108.8
1959	164.0	6.36	9.03	9.36	178.0	271	109.7
1960	170.0	6.58	9.17	10.36	178.0	270	113.1
1961	213.0	8.25	11.47	12.56	182.0	270	131.5
1962	214.0	8.25	11.47	12.56	182.0	270	136.5
1963	229.2	8.25	11.47	12.56	200.0	270	128.4
1964	229.2	8.25	11.47	12.56	200.0	240	123.7
1965	229.2	8.47	11.06	12.37	204.0	240	120.4
1966	236.2	9.81	13.43	14.83	204.0	240	120.0
1967	243.2	9.81	13.43	14.83	204.0	228	119.1
1968	241.2	9.81	13.43	14.83	204.0	228	119.2
1969	240.8	9.81	13.43	14.83	204.0	228	117.9
1970	241.2	9.81	13.43	14.83	204.0	228	117.6
1971	252.2	9.81	13.43	16.30	204.2	222	116.7
1972	256.0	9.81	13.43	16.30	210.0	222	116.5
1973	253.8	9.81	13.43	16.30	204.0	222	117.2
1974	252.0	9.81	13.43	16.30	210.0	226	117.8
1975	254.4	9.81	13.43	16.30	207.0	226	118.0
1976	255.6	9.81	13.43	16.30	201.6	236	118.3
1977	256.6	9.81	13.43	16.30	208.0	223	120.8
1978	263.4	9.81	13.61	20.06	227.8	231	121.6
1979	330.7	11.90	16.48	23.07	268.0	236	124.0
1980	360.6	11.97	16.31	23.06	317.4	237	131.4
1981	381.7	11.97	16.31	34.50	311.6	243	134.5
1982	392.2	11.97	16.31	34.50	323.6	260	137.1
1983	392.6	12.04	16.34	34.17	342.2	259	139.2
1984	395.1	12.04	16.43	31.05	341.8	322	143.1

Notes: Column (a) - Composite average price for milled grain products.
Column (b) - Standard price for mid-grade unmilled products.
Column (c) - Composite average price for ginned cotton products.
Column (d) - Composite average price for fertilizers in standard weights.
All the prices are offered of current prices by the official marketing system.
Column (e) - Composite average price index based on data collection for commodity prices from a certain number of cities and county towns.

Sources: For Columns (a), (c) and (d):
Figures up to 1982 are from *TJNJ 1983*, p477, 478
Figures for 1983, 1984 are from *TJNJ 1985*, pp546-548
For Column (b): Figures are from NJZL 1949-1983, pp492-493
For Column (e): Figures up to 1982 are from *TJNJ 1983*, p455
Figures for 1983 and 1984 are from *TJNJ 1985*, p530

A1.4 Regional Data of Pindu and Jianou Counties 1952-84

Year	Pindu		Jianou		Purchasing price (yuan/ tonne)
	Grain output (million jin)	Grain growing area (10,000 mu)	Grain output (million jin)	Grain growing area (10,000 mu)	
1952	568	418	137	60.9	125
1953	510	421	153	63.1	125
1954	522	415	160	64.2	124
1955	538	390	161	65.3	124
1956	613	388	175	76.7	124
1957	525	398	176	76.0	124
1958	564	397	174	69.6	124
1959	533	317	164	66.9	124
1960	317	379	158	80.4	124
1961	436	356	161	74.8	168
1962	401	351	166	75.1	168
1963	478	345	179	73.8	168
1964	422	340	191	77.5	168
1965	501	343	209	79.1	168
1966	640	329	176	78.9	196
1967	608	324	202	75.5	196
1968	544	321	185	72.9	196
1969	662	331	193	75.9	196
1970	656	327	267	78.2	196
1971	706	337	290	88.6	196
1972	704	333	301	94.2	196
1973	867	319	315	97.4	196
1974	761	319	345	98.5	196
1975	943	311	348	102.2	196
1976	930	308	314	100.0	196
1977	939	305	348	101.3	196
1978	1,089	299	385	98.5	196
1979	1,144	279	417	95.3	238
1980	1,186	273	424	93.0	238
1981	892	276	458	97.7	322
1982	1,207	267	475	95.5	331
1983	1,532	267	589	91.6	340
1984	1,588	266	498	89.2	342

Notes: Figures of grain output are for unmilled grain products.
Grain purchasing price of Jianou County is the standard purchasing price for unmilled indica rice product, which is the predominant grain crop in the county.

Source: Data is provided by the correspondent county planning commissions.

A1.5 Food Grain Consumption 1952-83 (in 100 million jin)

Year	Total of the nation	In which:		Year	Total of the nation	In which:	
		City & town	Country-side			City & town	Country-side
1952	2,244.1	333.7	1,910.4	1968	2,695.2	474.3	2,220.9
1953	2,286.5	357.3	1,929.2	1969	2,772.0	487.9	2,284.1
1954	2,333.1	379.2	1,953.9	1970	3,065.8	538.7	2,527.1
1955	2,408.5	359.0	2,049.5	1971	3,165.9	559.7	2,606.2
1956	2,534.0	378.0	2,156.2	1972	2,974.5	579.2	2,395.3
1957	2,583.5	392.0	2,191.5	1973	3,380.2	592.6	2,787.6
1958	2,584.8	432.8	2,152.0	1974	3,379.9	593.2	2,786.7
1959	2,471.8	522.8	1,949.0	1975	3,493.7	615.8	2,877.9
1960	2,172.2	532.5	1,638.7	1976	3,543.7	643.0	2,900.7
1961	2,093.4	465.8	1,627.6	1977	3,626.7	681.6	2,945.1
1962	2,188.2	426.2	1,762.0	1978	3,741.0	700.0	3,041.0
1963	2,242.9	427.8	1,815.1	1979	4,015.5	740.0	3,275.5
1964	2,536.9	467.9	2,069.0	1980	4,199.4	773.0	3,426.4
1965	2,630.6	516.3	2,114.3	1981	4,361.4	799.1	3,562.3
1966	2,797.0	492.3	2,304.7	1982	4,546.1	825.7	3,720.4
1967	2,812.8	495.1	2,317.7	1983	4,738.3	860.1	3,878.2

Note: Figures in this table are for milled grain.
Source: From the *ZCWTZ 1952 -1983*, p27.

A1.6 Per Capita Food Grain Consumption 1952-83 (in jin)

Year	Total of the nation	In which: City & town	In which: Country-side	Year	Total of the nation	In which: City & town	In which: Country-side
1952	395.34	480.70	383.44	1968	347.61	377.69	341.80
1953	394.14	484.47	380.99	1969	347.77	383.15	341.05
1954	392.75	472.41	380.31	1970	374.43	403.58	368.75
1955	396.53	428.81	391.37	1971	376.53	398.90	372.05
1956	408.58	400.51	410.03	1972	345.02	411.92	331.98
1957	406.12	392.00	408.76	1973	383.14	416.09	376.79
1958	396.46	371.09	401.99	1974	375.28	409.39	368.74
1959	373.18	401.78	366.19	1975	381.03	418.51	373.87
1960	327.24	385.17	311.99	1976	380.56	424.03	372.10
1961	317.57	358.97	307.42	1977	384.14	420.95	376.52
1962	329.25	367.67	321.13	1978	390.92	410.58	386.66
1963	329.29	379.71	319.30	1979	414.05	421.51	412.40
1964	363.91	400.09	356.62	1980	427.62	427.76	427.59
1965	365.67	421.30	354.25	1981	438.35	431.46	439.93
1966	379.14	411.52	372.87	1982	450.92	434.58	454.72
1967	372.36	398.50	367.22	1983	464.46	443.35	469.41

Note: Figures in this table are for milled grain.
Source: From the *ZCWTZ 1952 -1983*. p. 27.

A1.7 Per Capita Disposable Cash Income 1952-83 (in yuan)

Year	Average of the nation	In which: City & town	In which: Country-side	Year	Average of the nation	In which: City & town	In which: Country-side
1952	47.4	183.2	30.0	1968	84.1	306.3	45.8
1953	56.7	219.0	34.9	1969	85.4	322.5	47.5
1954	59.3	215.1	36.8	1970	86.7	314.6	48.9
1955	59.0	212.4	36.5	1971	91.3	326.4	52.0
1956	68.5	233.8	41.2	1972	97.7	363.0	53.6
1957	69.3	242.2	39.7	1973	102.1	378.5	56.6
1958	76.6	231.3	46.8	1974	105.3	390.6	57.6
1959	82.7	250.5	45.8	1975	110.1	401.5	60.1
1960	86.8	265.8	44.3	1976	113.9	409.9	61.2
1961	85.0	270.6	44.6	1977	120.4	418.1	65.2
1962	79.6	296.4	40.3	1978	131.0	440.8	70.4
1963	79.4	286.1	42.0	1979	157.0	486.1	92.2
1964	82.1	289.5	44.0	1980	193.5	569.1	118.3
1965	84.8	291.9	46.2	1981	210.2	585.1	134.6
1966	88.2	311.1	49.0	1982	228.7	609.8	149.3
1967	89.3	310.4	50.3	1983	258.7	637.5	175.1

Note: Figures in this table are in current prices.
Source: From the *ZCWTZ 1952 -1983*. p20.

A1.8 Total Population of the Nation, 1952-83 (in 10,000 persons)

Year	Total of the nation	In which:		Year	Total of the nation	In which:	
		City & town	Country-side			City & town	Country-side
1952	57,482	7,163	50,319	1968	78,534	13,838	64,696
1953	58,796	7,826	50,970	1969	80,671	14,117	66,554
1954	60,266	8,249	52,017	1970	82,992	14,424	68,568
1955	61,465	8,285	53,180	1971	85,229	14,711	70,518
1956	62,828	9,185	53,643	1972	87,177	14,935	72,242
1957	64,653	9,949	54,704	1973	89,211	15,345	73,866
1958	65,994	10,721	55,273	1974	90,859	15,595	75,264
1959	67,207	12,371	54,836	1975	92,420	16,030	76,390
1960	66,207	13,073	53,134	1976	93,717	16,341	77,376
1961	65,859	12,707	53,152	1977	94,974	16,669	78,305
1962	67,295	11,659	55,636	1978	96,259	17,245	79,014
1963	69,172	11,646	57,526	1979	97,542	18,495	79,047
1964	70,499	12,950	57,549	1980	98,705	19,140	79,565
1965	72,538	13,045	59,493	1981	100,072	20,171	79,901
1966	74,542	13,313	61,229	1982	101,541	21,154	80,387
1967	76,368	13,548	62,820	1983	102,495	24,126	78,369

Notes: The total population of the nation in this table covers the whole population of the 29 provinces, autonomous regions and metropolitan cities as well as the army of the People's Republic of China.

The population in city and town includes the whole population under the jurisdiction of city and township governments. The population in Countryside covers the whole population in counties except for those in towns.

All the population figures are for the year end population.

Sources: Figures for 1952-82 are from the *TJNJ 1983*, p 103
Figures for 1983 are from the *TJNJ 1985*, p 185

A1.9 Per Capita Cash Expenditure on Consumer Goods 1952-83 (in yuan)

Year	Average of the nation	In which:		Year	Average of the nation	In which:	
		City & town	Country-side			City & town	Country-side
1952	41.9	145.2	27.5	1968	76.7	257.2	41.9
1953	50.7	180.9	31.7	1969	79.6	263.5	44.6
1954	53.4	174.8	34.5	1970	81.4	257.6	47.1
1955	53.5	173.7	34.3	1971	82.4	255.9	47.6
1956	59.8	187.8	36.9	1972	87.6	287.2	48.6
1957	62.1	195.7	37.2	1973	91.8	302.8	51.2
1958	65.2	176.6	40.9	1974	95.4	317.5	52.8
1959	73.7	200.4	42.7	1975	100.6	336.2	55.6
1960	77.4	213.2	41.7	1976	103.7	346.2	56.5
1961	73.4	232.5	34.4	1977	109.8	349.8	60.1
1962	75.3	249.3	40.1	1978	117.2	364.5	63.5
1963	73.6	234.2	41.6	1979	134.9	381.2	80.4
1964	74.9	235.3	42.4	1980	163.2	433.8	102.2
1965	74.7	236.5	41.4	1981	180.0	453.9	117.4
1966	77.9	256.9	43.3	1982	193.2	465.6	129.9
1967	81.5	260.8	46.2	1983	211.8	486.1	147.3

Notes: All the figures in this table are in current prices.

Figures for average of the nation and countryside do not include the peasant household's consumer goods expenditure in kind, either goods distributed by the basic production units or from the household's sideline occupation.

Source: From the *ZCWTZ 1952-1983*, p21

A1.10 Grain Products for Feed, Seed and Industrial Use 1952-83 (in 100 million jin)

Year	Grain for feed	Grain for seed	Grain for industrial use	Year	Grain for feed	Grain for seed	Grain for industrial use
1952	195.8	147.8	36.8	1968	388.2	315.2	44.1
1953	191.6	154.0	39.3	1969	380.5	338.2	46.1
1954	195.9	158.4	40.9	1970	421.1	363.0	51.6
1955	228.9	173.1	40.9	1971	472.5	384.5	56.9
1956	286.5	219.7	44.3	1972	489.1	404.9	58.8
1957	245.8	225.5	25.8	1973	550.6	420.9	58.5
1958	267.1	291.7	28.6	1974	581.7	440.8	61.0
1959	214.3	263.8	45.3	1975	627.9	477.9	65.9
1960	178.2	242.5	33.7	1976	665.0	485.0	68.2
1961	161.8	229.9	17.9	1977	662.0	479.0	70.5
1962	198.5	234.1	24.8	1978	745.0	481.0	72.8
1963	232.4	241.1	27.1	1979	800.0	485.0	73.2
1964	258.0	249.2	30.5	1980	878.0	470.0	100.3
1965	375.3	276.0	36.4	1981	884.2	463.1	115.3
1966	404.6	295.7	41.2	1982	910.4	458.7	124.2
1967	402.4	305.4	42.5	1983	882.2	469.3	142.8

Notes: Grain for feed refers to grains for feeding animals (in China's case this involves pigs, cattle, horses, donkeys, mules, camels, sheep and goats) and poultry (chicken, duck and geese). It covers both feed grain distributed by the official marketing system and that retained by peasant households and basic production units. It is derived by multiplying the number of individual animals and poultry with correspondent unit feeding standards and then summing up. Grain for seed covers both seeds distributed by the official marketing system and that retained by agricultural producers. It is derived by multiplying annual growing areas of individual cereal crops with correspondent unit seeding standards and then summing up. Grain for industrial use refers to grains used in industrial and handicraft sectors, for making cake, biscuits, wine, ethyl alcohol, soy sauce, vinegar and sizing, etc. All the figures in this table are for milled grain.

Source: From China's Statistical Information and Consultancy Service Centre, State Statistical Bureau.

A1.11 Total Consumption of Wine, Biscuit and Cake 1957-83 (in 10,000 tonnes)

Year	Wine	Biscuit and cake	Year	Wine	Biscuit and cake
1957	86.7	50.4	1971	139.9	120.9
1958	92.7	58.5	1972	157.4	140.0
1959	122.3	72.0	1973	163.0	142.3
1960	128.2	69.0	1974	179.6	161.7
1961	73.3	70.0	1975	199.5	164.2
1962	75.9	74.8	1976	232.8	168.5
1963	81.4	100.0	1977	236.7	182.3
1964	86.0	87.5	1978	246.0	184.8
1965	93.7	81.8	1979	289.2	181.7
1966	99.4	74.5	1980	334.2	201.0
1967	108.2	82.7	1981	439.2	202.0
1968	98.9	83.6	1982	528.8	201.3
1969	115.0	119.2	1983	592.7	208.6
1970	123.3	120.0			

Notes: The wine for human consumption in China refers to spirits, beer, grape wine and yellow and millet wines. Figures for the total consumption of wine covers wine sold to the consumers both by the industrial and commercial sectors. It does not include either wine made by consumers themselves or that used by the drug making industry. Figures for the total consumption of biscuit and cake refers to bread, biscuit, cake, pastry and various Chinese and western sandwiches sold by the official retailing shops.

Source: From the *ZCWTZ 1952-1983*, pp34, 51

A1.12 Retail Prices of the Grain Products 1952-83

Year	Average retail price (yuan/ tonne)	Retail price index in city & town (1952= 100)	Retail price index in the coun- tryside (1952= 100)	Year	Average retail price (yuan/ tonne)	Retail price index in city & town (1952= 100)	Retail price index in the coun- tryside (1952= 100)
1952	197.8	100.0	100.0	1968	260.0	122.3	136.7
1953	201.6	105.1	109.2	1969	260.0	122.3	136.7
1954	205.0	105.1	109.3	1970	260.0	122.3	136.7
1955	212.6	105.4	109.3	1971	260.0	122.3	136.8
1956	212.6	105.4	109.3	1972	277.6	122.3	136.8
1957	220.0	106.1	108.8	1973	277.6	122.4	136.8
1958	220.0	106.1	109.1	1974	287.2	122.4	136.8
1959	220.0	106.3	109.7	1975	288.0	122.4	136.8
1960	227.2	107.1	111.3	1976	288.2	122.4	136.8
1961	231.0	107.4	111.2	1977	292.0	122.4	136.8
1962	229.4	107.4	111.3	1978	294.8	122.4	136.8
1963	230.0	108.2	112.6	1979	298.6	122.4	141.0
1964	230.0	108.2	113.9	1980	307.6	122.4	142.3
1965	237.4	112.5	122.0	1981	337.1	337.1	142.4
1966	245.6	117.3	127.2	1982	340.3	340.3	142.7
1967	257.2	122.3	136.7	1983	351.4	351.4	142.8

Notes: Average retail price of grain products is the composite average price after taking account of the differences in standard, grade and quality of grain products. It is derived by dividing the amount of grain sold from the total retail value.

Retail price index in city and town and in countryside are the prices from official marketing agency, take 1952 as 100.

Sources: Figures for the average retail price are from *TJNJ 1983*, p470 and *TJNJ 1985*, p543.

Figures for the retail price index in city and town and in countryside are from *ZCWTZ 1952-1983*, pp382, 388.

A1.13 Total Consumption of Pork and Fresh Eggs 1957-83

Year	Pork (in 100 million jin)	Fresh eggs (in million jin)	Year	Pork (in 100 million jin)	Fresh eggs (in million jin)
1952	67.1	1,158.9	1968	101.8	1,760.8
1953	70.3	1,243.4	1969	94.1	1,912.4
1954	71.4	1,300.1	1970	98.5	2,158.9
1955	60.0	1,272.5	1971	118.1	2,723.0
1956	57.8	1,470.8	1972	130.3	2,638.8
1957	64.6	1,597.7	1973	134.6	2,864.0
1958	68.1	1,660.3	1974	138.1	2,838.1
1959	40.8	766.2	1975	139.8	2,989.1
1960	20.3	644.1	1976	137.4	3,282.1
1961	18.6	655.5	1977	136.8	3,491.5
1962	29.5	1,019.0	1978	146.8	3,766.2
1963	58.2	1,404.0	1979	187.4	4,025.8
1964	78.3	1,876.5	1980	219.1	4,458.2
1965	90.4	2,043.5	1981	220.4	4,850.0
1966	103.9	2,532.0	1982	237.0	5,089.4
1967	104.0	2,314.2	1983	251.9	6,040.1

Note: Figures for pork refer to the weight of slaughtered carcass.

Source: From *ZCWTZ 1952-1983*, pp29, 30, 50.

A1.14 The Purchasing Price Index of Grain and Livestock Products 1952-83

Year	Grain	Livestock products	In which: Pork	Year	Grain	Livestock products	In which: Pork
1952	100.0	100.0	100.0	1968	190.0	183.3	188.1
1953	112.9	110.0	112.7	1969	189.1	183.3	188.1
1954	112.9	117.7	120.5	1970	187.2	183.3	188.1
1955	113.1	116.4	117.1	1971	194.5	183.5	189.6
1956	115.2	120.4	122.2	1972	197.0	185.1	190.7
1957	116.5	137.7	139.1	1973	195.8	186.1	191.5
1958	119.5	142.2	143.6	1974	199.0	189.0	193.8
1959	121.1	146.2	144.7	1975	201.3	189.8	194.2
1960	125.0	154.0	148.8	1976	213.4	198.8	194.2
1961	216.6	183.7	178.9	1977	211.7	190.0	194.4
1962	228.8	181.2	182.4	1978	213.3	192.0	195.0
1963	190.4	182.5	183.8	1979	278.3	236.9	242.2
1964	170.4	182.8	188.1	1980	301.7	248.2	248.7
1965	170.4	181.7	188.1	1981	331.0	252.2	249.4
1966	192.2	182.1	188.1	1982	343.6	253.2	250.1
1967	191.5	183.3	188.1	1983	379.0	254.2	249.8

Notes: Figures in this table are the weighted average price index, take 1952 as 100.

Source: From *ZCWTZ 1952-1983,* pp402, 411.

A1.15 Trade of Grain Products 1961-83 (in 10,000 tonnes)

Year	Rice export	Wheat import	Net grain import
1961	42.83	388.17	445.48
1962	45.79	353.56	389.21
1963	68.45	558.77	446.19
1964	76.16	536.87	474.93
1965	98.49	607.27	398.88
1966	148.74	621.38	355.28
1967	157.66	439.46	170.75
1968	129.92	445.14	199.50
1969	117.85	374.02	154.88
1970	127.95	530.21	324.05
1971	129.20	302.20	55.97
1972	142.57	433.36	183.07
1973	263.08	629.85	423.48
1974	206.06	538.34	447.74
1975	162.96	349.12	92.89
1976	87.61	202.19	60.17
1977	103.29	687.58	568.78
1978	143.52	766.73	695.53
1979	105.31	870.98	1,070.45
1980	111.64	1,097.17	1,181.10
1981	58.33	1,307.01	1,355.14
1982	45.71	1,353.43	1,485.57
1983	56.59	1,101.91	1,147.20

Notes: Figures for rice export are for milled rice. All the figures presented are from the foreign trade administrative department and not that of the Customs.

Source: From *ZCWTZ 1952-1983*, pp496, 510 and 516.

A1.16 Trading Price of Rice, Wheat and Fertilizer 1961–83

Year	Rice export ($/tonne)	Wheat import ($/tonne)	Fertilizer import ($/tonne)
1961	109	67	40
1962	116	75	38
1963	112	66	37
1964	116	77	40
1965	120	71	52
1966	126	71	49
1967	138	75	33
1968	147	70	37
1969	131	58	36
1970	118	51	28
1971	103	61	25
1972	105	63	28
1973	274	102	40
1974	405	151	61
1975	338	152	102
1976	275	137	62
1977	247	95	57
1978	308	107	71
1979	321	129	85
1980	350	167	117
1981	411	179	129
1982	382	179	101
1983	304	146	84

Notes: Prices of rice export are for milled rice in free on board (f.o.b.)
Prices of wheat and fertilizer are for the cost, insurance and freight (c.i.f.)
The price figures are derived from dividing the value of exports or imports by the volume of corresponding trading goods.

Source: From *ZCWTZ 1952-1983*, pp496, 510 and 512.

Appendix Two

Simple Correlation Matrices

Matrix 2.1 (corresponding to variables in Table 4.4)

	GOUT	GAREA	WEATH	GPURP	TIME	PRATOGC	PRATOGF*
GOUT*	1.00	−0.57	0.09	0.90	0.92	0.55	0.83
GAREA	−0.57	1.00	−0.11	−0.75	−0.73	−0.64	−0.77
WEATH	0.09	−0.11	1.00	0.25	0.19	0.10	0.18
GPURP	0.90	−0.75	0.25	1.00	0.93	0.68	0.93
TIME	0.92	−0.73	0.19	0.93	1.00	0.79	0.95
PRATOGC	0.55	−0.64	0.10	0.68	0.79	1.00	0.83
PRATOGF	0.83	−0.77	0.18	0.93	0.95	0.83	1.00

Matrix 2.2 (corresponding to variables in Table 4.6)

	ROUT	RAREA	WEATH	RPURP	TIME	PRATORC	PRATORF
ROUT	1.00	0.76	0.05	0.86	0.93	0.13	0.74
RAREA	0.76	1.00	−0.08	0.56	0.70	0.31	0.60
WEATH	0.05	−0.08	1.00	0.14	0.19	−0.17	0.06
RPURP	0.86	0.56	0.14	1.00	0.95	0.45	0.94
TIME	0.93	0.70	0.19	0.95	1.00	0.38	0.90
PRATORC	0.13	0.31	−0.17	0.45	0.38	1.00	0.66
PRATORF	0.74	0.60	0.06	0.94	0.90	0.66	1.00

Matrix 2.3 (corresponding to variables in Table 4.7)

	WOUT	WAREA	WEATH	WPURP	TIME	PRATOWC	PRATOWF
WOUT	1.00	0.79	0.20	0.81	0.87	−0.14	0.62
WAREA	0.79	1.00	0.35	0.51	0.62	−0.23	0.35
WEATH	0.20	0.35	1.00	0.14	0.19	−0.22	0.06
WPURP	0.81	0.51	0.14	1.00	0.95	0.32	0.93
TIME	0.87	0.62	0.19	0.95	1.00	0.26	0.90
PRATOWC	−0.14	−0.23	−0.22	0.32	0.26	1.00	0.56
PRATOWF	0.62	0.35	0.06	0.93	0.90	0.56	1.00

Note: *See p356 for definitions of these variables

Matrix 2.4 (corresponding to variables in Table 4.8)

	BOUT	BAREA	WEATH	BPURP	TIME	PRATOBC	PRATOBF*
BOUT*	1.00	0.45	−0.33	0.11	−0.14	−0.14	−0.01
BAREA	0.45	1.00	−0.06	−0.62	−0.87	−0.79	−0.73
WEATH	−0.33	−0.06	1.00	0.27	0.19	0.23	0.23
BPURP	0.11	−0.62	0.27	1.00	0.88	0.92	0.97
TIME	−0.14	−0.87	0.19	0.88	1.00	0.93	0.93
PRATOBC	−0.14	−0.79	0.23	0.92	0.93	1.00	0.96
PRATOBF	−0.01	−0.73	0.23	0.97	0.93	0.96	1.00

Matrix 2.5 (corresponding to variables in Table 4.10)

	OTOUT	OTAREA	WEATH	GPURP	TIME	PRATOGC	PRATOGF
OTOUT	1.00	−0.87	0.08	0.88	0.92	0.60	0.84
OTAREA	−0.87	1.00	−0.14	−0.90	−0.94	−0.63	−0.88
WEATH	0.08	−0.14	1.00	0.25	0.19	0.10	0.18
GPURP	0.88	−0.90	0.25	1.00	0.93	0.68	0.93
TIME	0.92	−0.94	0.19	0.93	1.00	0.79	0.95
PRATOGC	0.60	−0.63	0.10	0.68	0.79	1.00	0.83
PRATOGF	0.84	−0.88	0.18	0.93	0.95	0.83	1.00

Matrix 2.6 (corresponding to variables in Table 4.12)

	GOUT	GAREA	WEATH	GPURP	TIME	PRATOGC	PRATOGF
GOUT	1.00	−0.07	0.14	0.84	0.84	0.42	0.73
GAREA	0.07	1.00	−0.07	−0.09	0.11	0.18	0.00
WEATH	0.14	−0.07	1.00	0.25	0.19	0.10	0.18
GPURP	0.84	−0.09	0.25	1.00	0.93	0.68	0.93
TIME	0.84	0.11	0.19	0.93	1.00	0.79	0.95
PRATOGC	0.42	0.18	0.10	0.68	0.79	1.00	0.83
PRATOGF	0.73	−0.01	0.18	0.93	0.95	0.83	1.00

Matrix 2.7 (corresponding to variables in Table 4.13)

	GOUT	GAREA	WEATH	TIME	PRATOGC	PRATOGF	GPURP2*
GOUT*	1.00	0.82	0.22	0.93	0.59	0.86	0.89
GAREA	0.82	1.00	0.25	0.87	0.63	0.76	0.63
WEATH	0.22	0.25	1.00	0.19	−0.04	0.12	0.18
TIME	0.93	0.87	0.19	1.00	0.76	0.94	0.88
PRATOGC	0.59	0.63	−0.04	0.76	1.00	0.85	0.75
PRATOGF	0.86	0.76	0.12	0.94	0.85	1.00	0.93
GPURP2	0.89	0.63	0.18	0.88	0.75	0.93	1.00

Note: * See pp356-357 for definitions of these variables.

Matrix 2.8 (corresponding to variables in Table 6.3)

	GOUT	CTFGDM	CTPOPTON	CTCHIM	CTBFEP	SPFGRN	PCGT
GOUT	1.00	0.92	0.90	0.93	0.93	0.79	0.85
CTFGDM	0.92	1.00	0.97	0.96	0.96	0.76	0.63
CTPOPTON	0.90	0.97	1.00	0.96	0.96	0.81	0.59
CTCHIM	0.93	0.96	0.96	1.00	0.99	0.74	0.69
CTBFEP	0.93	0.96	0.96	0.99	1.00	0.78	0.66
SPFGRN	0.79	0.76	0.81	0.74	0.78	1.00	0.48
PCGT	0.85	0.63	0.59	0.69	0.66	0.48	1.00

Matrix 2.9 (corresponding to variables in Table 6.4)

	GOUT	CSFGDM	CSPOPTON	CSCHIM	CSBFEP	SPFGRN	PCGT	PPFGRN
GOUT	1.00	0.99	0.93	0.86	0.86	0.85	0.85	0.82
CSFGDM	0.99	1.00	0.91	0.88	0.89	0.82	0.87	0.82
CSPOPTON	0.93	0.91	1.00	0.71	0.73	0.93	0.65	0.78
CSCHIM	0.86	0.88	0.71	1.00	0.99	0.63	0.73	0.90
CSBFEP	0.86	0.89	0.73	0.99	1.00	0.65	0.73	0.90
SPFGRN	0.85	0.82	0.93	0.63	0.65	1.00	0.55	0.75
PCGT	0.85	0.87	0.65	0.73	0.73	0.55	1.00	0.56
PPFGRN	0.82	0.82	0.78	0.90	0.90	0.75	0.56	1.00

Matrix 2.10 (corresponding to variables in Table 6.5)

	GOUT	GCSPTON	POPTON	PACHIM	SPFGRN*
GOUT*	1.00	0.99	0.94	0.88	0.94
GCSPTON	0.99	1.00	0.94	0.92	0.95
POPTON	0.94	0.94	1.00	0.83	0.97
PACHIM	0.88	0.92	0.83	1.00	0.91
SPFGRN	0.94	0.95	0.97	0.91	1.00

Matrix 2.11 (corresponding to variables in Table 6.6)

	FDGRAN	TCPORK	TCEGGS	PRTAPG	PRTPKG	TIME
FDGRAN	1.00	0.96	0.97	−0.70	−0.69	0.94
TCPORK	0.96	1.00	0.98	−0.70	−0.71	0.88
TCEGGS	0.97	0.98	1.00	−0.72	−0.72	0.90
PRTAPG	−0.70	−0.70	−0.72	1.00	0.98	−0.70
PRTPKG	−0.69	−0.71	−0.72	0.98	1.00	−0.66
TIME	0.94	0.88	0.90	−0.70	−0.66	1.00

Note: * See p357 for definitions of these variables

Matrix 2.12 (corresponding to variables in Table 6.7)

	SDGRAN	GAREA	TIME
SDGRAN	1.00	−0.59	0.96
GAREA	−0.59	1.00	−0.71
TIME	0.96	−0.71	1.00

Matrix 2.13 (corresponding to variables in Table 6.8)

	DGRAN	TCBCCK	TCWINCE	TIME
IDGRAN	1.00	−0.52	−0.22	0.89
TCBCCK	−0.52	1.00	0.21	−0.37
TWCINCE	−0.22	0.21	1.00	−0.24
TIME	0.89	−0.37	−0.24	1.00

Matrix 2.14 (corresponding to variables in Table 7.3)

	REPORT	REPRCE	GCSPTON	ROUT	PRATORW	PRATORF*
REPORT*	1.00	0.10	−0.02	0.11	0.46	0.71
REPRCE	0.10	1.00	0.84	0.80	0.67	0.37
GCSPTON	−0.02	0.84	1.00	0.97	0.54	0.22
ROUT	0.11	0.80	0.97	1.00	0.59	0.33
PRATORW	0.46	0.67	0.54	0.59	1.00	0.65
PRATORF	0.71	0.37	0.22	0.33	0.65	1.00

Matrix 2.15 (corresponding to variables in Table 7.4)

	WIPORT	WIPRCE	GCSPTON	GOUT	PRATOWR	PRATOWF
WIPORT	1.00	0.68	0.76	0.69	−0.34	−0.42
WIPRCE	0.68	1.00	0.83	0.81	−0.47	−0.16
GCSPTON	0.76	0.83	1.00	0.99	−0.59	−0.15
GOUT	0.69	0.81	0.99	1.00	−0.61	−0.10
PRATOWR	−0.34	−0.47	−0.59	−0.61	1.00	0.07
PRATOWF	−0.42	−0.16	−0.15	−0.10	0.07	1.00

Matrix 2.16 (corresponding to variables in Table 7.5)

	GRIMPT	WIPRCE	GCSPTON	GOUT	REPRCE	PRATORW
GRIMPT	1.00	0.69	0.72	0.66	0.65	0.27
WIPRCE	0.69	1.00	0.83	0.81	0.95	0.43
GCSPTON	0.72	0.83	1.00	0.99	0.84	0.54
GOUT	0.66	0.81	0.99	1.00	0.83	0.56
REPRCE	0.65	0.95	0.84	0.83	1.00	0.67
PRATORW	0.27	0.43	0.54	0.56	0.67	0.00

Note: *See p358 for definitions of these variables

Notes to Accompanying Correlation Matrices

Matrix 2.1

GOUT	=	Total unmilled grain output in 0.1 million tonnes
GAREA	=	Grain growing area in 0.1 million hectares
WEATH	=	Crop growing area suffered from natural calamities with 30% or more reduction in yield, measured in 10,000 hectares
GPURP	=	Grain purchasing price in yuan/ tonne, 2.8 yuan = $1
TIME	=	Time trend variable for technological improvement, 1952-84
PRATOGC	=	Price ratio of grain over cotton, both are governmental purchasing prices
PRATOGF	=	Price ratio of purchasing price of grain over selling price of fertilizer

Matrix 2.2

ROUT	=	Total unmilled rice product in 0.1 million tonnes
RAREA	=	Rice growing area in 10,000 hectares
RPURP	=	Rice purchasing price in yuan/ tonne
PRATORC	=	Price ratio of rice over cotton
PRATORF	=	Price ratio of rice over fertilizer

Matrix 2.3

WOUT	=	Total unmilled wheat products in 10,000 tonnes
WAREA	=	Wheat growing area in 10,000 hectares
WPURP	=	Wheat purchasing price in yuan/ tonne
PRATOWC	=	Price ratio of wheat over cotton
PRATOWF	=	Price ratio of wheat over fertilizer

Matrix 2.4

BOUT	=	Soybean output in 10,000 tonnes
BAREA	=	Soybean growing area in 10,000 hectares
BPURP	=	Soybean purchasing price in yuan/ tonne
PRATOBC	=	Price ratio of soybean over cotton
PRATOBF	=	Price ratio of soybean over fertilizer

Matrix 2.5

OTOUT	=	Other grain products (except from rice, wheat and soybean) in million tonnes
OTAREA	=	Growing area of other grain crops in 10,000 hectares

Matrix 2.6

GOUT	=	Total grain output in 1,000 tonnes
GAREA	=	Grain growing area in 100 hectares

Matrix 2.7

GOUT	=	Total grain output in 1,000 tonnes
GAREA	=	Grain growing area in 1,000 hectares
GPURP2	=	Grain purchasing price of the county in yuan/ tonne

Matrix 2.8

CTFGDM = Total food grain consumption in city and towns in 10,000 tonnes (milled)
CTPOPTON = Total population in city and towns in 0.1 million persons
CTCHIM = Per capita disposable income for citizens in city and towns in yuan
CTBFEP = Per capita consumer goods expenditure for citizens in city and towns in yuan
SPFGRN = Food grain retail price index in city and towns, take 1952 as 100
PCGT = Per capita grain availability of the nation on average

Matrix 2.9

CSFGDM = Food grain consumption of the countryside in 0.1 million tonnes
CSPOPTON = Total rural population in million persons
CSCHIM = Per capita disposable cash income of rural population in yuan
CSBFEP = Per capita consumer goods expenditure of rural population in yuan
PPRGRN = Price index of grain purchasing (weighted average of quota purchasing, over-quota premium and purchasing with negotiated price), take 1952 as 100

Matrix 2.10

GCSPTON = Milled grain for domestic food consumption in 0.1 million tonnes
POPTON = Total population of the nation in million persons
PACHIM = National per capita disposable income in yuan
SPFGRN = Retail price of food grain products in yuan

Matrix 2.11

FDGRAN = Feed grain consumption in 10,000 tonnes
TCPORK = Total pork consumption in 10,000 tonnes
TCEGGS = Total fresh eggs consumption in 10,000 tonnes
PRTAPG = Price ratio of animal products to grain
PRTPKG = Price ratio of pork to grain

Matrix 2.12

SDGRAN = Demand of seed grain in 10,000 tonnes

Matrix 2.13

IDGRAN = Grain for industrial use in 10,000 tonnes (milled)
TCBCCK = Total consumption of biscuit and cake in 10,000 tonnes
TCWINE = Total consumption of wine in 10,000 tonnes

Matrix 2.14

REPORT = Rice export in 10,000 tonnes
REPRCE = Rice export price in $/tonne
PRATORW = Price ratio of rice to wheat
PRATORF = Price ratio of rice to fertilizer

Matrix 2.15

WIPORT	=	Wheat import in 10,000 tonnes
WIPRICE	=	Wheat import price in $/tonne
PRATOWR	=	Price ratio of wheat to rice
PRATOWF	=	Price ratio of wheat to fertilizer

Matrix 2.16

GRIMPT	=	Net grain import in 10,000 tonnes

References

Abbott, J.C. (1977) Food and Agricultural Marketing in China, *Food Policy*, November 1977, pp318–330.

Agricultural Publishing House ed. (1983) *Agriculture in China*, Beijing, China.

Agricultural Statistical Bureau of the State Statistical Bureau ed. (1986) *China's Rural Statistical Yearbook 1985* (*Zhongguo Nongchun Tongji Nianjian 1985*), Chinese Statistical Press, Beijing, China.

Akiyama, T. and Mitchell, D.O. (1988) *Outlook for Beverages, Grains and Soybeans to 2000*. A research paper presented at the XX International Conference of Agricultural Economists, Buenos Aires, Argentina, August 26 – September 2, 1988.

An Pingsheng (1983) On the Development of the Agricultural Responsibility Systems, *Economic Management* (*Jingji Guanli*), March 1983, pp5–7.

Aziz, S. (1978) *Rural Development –Learning from China,* Macmillan Press Ltd., London and Basingstoke.

Barker, Sinha and Rose ed. (1982) *The Chinese Agricultural Economy*, Westview Press/ Croom Helm.

Barnett, A.D. (1981) *China's Economy in Global Perspective*, The Brookings Institution, Washington, D.C.

Beach, C.M. and Mackinnon, J.G. (1978) A Maximum Likelihood Procedure for Regression with Auto-correlated Error, *Econometrics* 46, pp51–58.

Beijing Review (1987) Party Congress Highlights Reform, Vol. 30, No. 44, Nov. 2–8, 1987, pp5–8.

Booth, A. (1986) *Food Trade and Food Security in Asia and Australia,* Joint Research Project, Kuala Lumpur and Canberra.

Buck, J.L. (1930) *China's Farm Economy*, Chicago, University of Chicago Press.

Buck, J.L. (1937) *Land Utilization in China*, Nanking, University of Nanking.

Bureau of Planning, Chinese Ministry of Agriculture (1981) *Statistical Collections of Foreign Agriculture and Animal Husbandry (1978–1979)*, Agricultural Press of China.

CAAS (1985) Abstract of the Comprehensive Report on Study of the Development of Grain and Cash Crops Production in China, in Vol. 4 of *Study of the Development of Grain and Cash Crops Production in China*, Edited by the Grain and Cash Crops Development Research Group, CAAS, June 1985, pp5–16.

Carter, C.A. and Zhong Fu-Ning, (1988) *China's Grain Production and Trade*, Westview Press, Boulder and London.

Chao Juo-Chun (1960) *Agrarian Policy of the Chinese Communist Party*, Asia Publishing House.

Chen, C.S. and Ridley, C.P. (1969) *Rural People's Communes in Lien-Chiang. Documents concerning Communes in Lien-Chiang County, Fukien Province, 1962-63*, Hoover Institution Publications, Stanford.

Chen Liangyu (1983a) Re-study on the Price-scissors between Industrial and Agricultural Products, *Study of Rural Development (Nongchun Huazhan Tansuo)*, No. 5, pp83–97.

Chen Liangyu (1983b) Some problems which should be given attention in making Agro-development Programme, *Agrotechnical Economy (Nongye Jishu Jingji)*, April 1983, pp11-15.

Chen Ting-Chung (1978) Planned Marketing by the State: Economic Fetters in Mainland China, *Issues and Studies (Wenti yu Yanjiu)*, January 1978, Vol. 14, No.1, pp28–39.

Chen Xikang, Chen Liangyu and Xie Xinwei (1983) Model of Optimal Farming Structure and its Applications, *Agroeconomy (Nongye Jingji)*, January 1983.

Chen Xikang, Chen Liangyu and Xie Xinwei (1985) Input-Output Analysis and its Application in Chinese Agricultural Sector, *Agricultural Investment Effectiveness (Nongye Touzhi Hiaoguo)*, 1, 1985.

Chen Yun (1951) The Development of Agriculture is of Paramount Importance, in *Collections of Chen Yun's Works 1949–1956* (1984) (in Chinese), The People's Press, pp140–143.

Chen Yun (1952) The Principle and Tasks of the Financial and Economic Aspects in 1952, in *Collections of Chen Yun's Works 1949–1956* (1984) (in Chinese), The People's Press, pp157–63.

Chen Yun (1953a) On the Implementation of Unique Purchasing and Selling of Grain Products, in *Collections of Chen Yun's Works 1949–1956* (1984) (in Chinese), The People's Press, pp202–216.

Chen Yun (1953b) The Production and Marketing Situations of the Edible Oils and its Solution, in *Collections of Chen Yun's Works 1949–1956* (1984) (in Chinese), The People's Press, pp217–221.

Chen Yun (1953c) On the Improvement of the Production and Marketing Conditions of Hu-shi-Ping, in *Collections of Chen Yun's Works 1949–1956* (1984) (in Chinese), The People's Press, pp222–228.

Chen Yun (1954) Questions concerning the Central Purchase and Supply, in *Collections of Chen Yun's Works 1949–1956* (1984) (in Chinese), The People's Press, pp254–263.

Chen Yun (1955) On Stick to and Improvement of the State Monopoly for Purchasing and Marketing of Grain Products, in *Collections of Chen Yun's Works 1949–1956* (1984) (in Chinese), The People's Press, pp272–279.

Chen Yun (1956) Methods for Improving the Supply Conditions of Pork and Other Hu-Shi-Ping, in *Collections of Chen Yun's Works 1956–1985* (1986) (in Chinese), The People's Press, pp16–27.

Chen Yun (1957a) On the Regulations for Improving the Commercial Management System, in *Collections of Chen Yun's Works 1956–1985* (1986) (in Chinese), The People's Press, pp86–89.

Chen Yun (1957b) Attention Must be Paid on Grain Purchasing and Distribution, in *Collections of Chen Yun's Works 1956–1985* (1986) (in Chinese), The People's Press, pp60–65.

Chen Yun (1961) On the Improvement of the Nation's Foreign Trade, in *Collections of Chen Yun's Works 1956–1985* (1986) (in Chinese), The People's Press, pp146–150.

Chen Zhiping (1985) Price Reform Key to Increasing Production, *Beijing Review*, Vol. 28, No. 16, pp15–16, 21.

Chia Hsiu-yen (1979) Price Problems in China's Economic Readjustment and Reform, *Bulletin of Nankai University*, No. 4, 1979, pp15–19.

China Handbook Editorial Committee (1984) *Economy*, China Handbook Series, Foreign Languages Press, Beijing, China.

Chinese Communist Party Central Committee (1978) Decisions on some problems in Accelerating Agricultural Development (draft), in *Research on Chinese Communism (Chung-Kung Yeu-Chiu)*, Vol. 13, No. 5, pp149–162.

Chinese Ministry of Agriculture, Animal Husbandry and Fishery ed. (1984) *Statistics of Agricultural Economy 1949–1983 (Nongye Jingji Ziliao 1949–1983)*, Chinese Agricultural Press, Beijing, China.

Chu Chin-Chih (1958) *China's Grain Policy and the Supply of Grain to Cities and Towns* (in Chinese).

Coale, A.J. and Hoover, E.M. (1958) *Population Growth and Economic Development in Low-income Countries: A Case Study of India's Prospects*, Princeton University Press, Princeton, New Jersey.

Colman, D.R. (1972) *The United Kingdom Cereal Market: An Econometric Investigation into the Effects of Pricing Policies*, Manchester University Press, Manchester.

Colman, D.R. (1983) A Review of the Arts of Supply Response Analysis, *Review of Marketing and Agricultural Economics*, Vol. 51, No. 3.

Coordinating Office of the National Agricultural Zoning Committee of China (1985) Several Strategic Issues of China's Agricultural Development, in *On the Development Strategy of Chinese Countryside*, Lu Wen, Niu Ruo Feng & Hu Yi Zun eds, Chinese Press of Agricultural Science and Technology, pp490–501.

Cowling, K. and Gardner, T.W. (1963) Analytical models for estimating supply relations in the agricultural sector: A survey and critique, *J. Agr. Econ.* 16 (2).

Dernberger, R.F. (1982) Agriculture in Communist Development Strategy, in *The Chinese Agricultural Economy* , Barker, Sinha and Rose eds, Westview Press/Croom Helm.

Donnithorne, A. (1970) *China's Grain: Output, Procurement, Transfers and Trade*, The Chinese University of Hong Kong, Economic Research Centre.

Dou Cuiping ed. (1984), *Agricultural Zoning of Pindu County (Pindu Xian Nongye Quhua)*, Agricultural Zoning Office of Pindu County.

Du Rensheng (1984) China's Countryside under Reform, *Beijing Review*, Vol. 27, No. 33, August 13, 1984.

Du Rensheng (1985) Second-Stage Rural Structural Reform, *Beijing Review*, Vol. 28, No. 25, June 24, 1985, pp15–17, 22.

Du Qiang (1984) Approach on the Solutions for the Problem of Rural Labour Surplus, *Bulletin of Jiangxi Financial and Economic College* No. 4, 1984, pp129–133.

Eckstein, A. (1961) The Strategy of Economic Development in Communist China, *American Economic Review*, Vol. 51, pp 508–517.

Fei, J.C.F. and Chiang, A. (1966) Growth under Austerity, in *The Theory and Design of Economic Development*, Irma Adelman and Eric Thorbecke eds, Johns Hopkins University Press, Baltimore.

Fisher, B.S. and Tanner, C. (1978) The Formulation of Price Expectations: An Empirical Test of Theoretical Models, *Amer. J. Agr. Econ.* 60: pp245–248.

FAO (1971) *Analysis of Supply Response to Price Changes*, Projections Research Working Paper No. 7, Rome.

FAO (1985) *Fertilizer Year Book*, Rome.

FAO (1987) *Agriculture: Toward 2000*, Rome.

Filtzer, D.A. ed. (1980), *The Crisis of Soviet Industrialization*, Selected Essays of E.A. Preobrazhensky, Macmillan.

Forester-Walker, K. (1987) China frees farming from politics, *New Scientist*, No. 1560, May 14, 1987, pp41–44.

Gek-boo Ng (1979) Incentive Policy in Chinese Collective Agriculture, *Food Policy*, May 1979, pp75–86.

Grain and Cash Crops Development Research Group of the Chinese Academy of Agricultural Science (1985) Approach on the Development of Chinese Grain and Cash Crops, in *On the Development Strategy of Chinese Countryside*, Lu Wen, Niu Ruo Feng, & Hu Yi Zun, eds, Chinese Press of Agricultural Science and Technology.

Greer, C. (1979) *Water Management in the Yellow River Basin of China*, University of Texas Press, U.S.A.

Griffin, K. and Saith, A. (1981) *Growth and Equality in Rural China*, Asian Employment Programme, The Asian Regional Team for Employment Promotion.

Gong Hongbao, (1984) Analysis of the current grain consumption conditions in China and projection for the year 2000, in *Study of the Development of Grain and Cash Crop Production in China*, Vol. 1. Edited by the Grain and Cash Crops Development Research Group, CAAS, March 1984, pp135–140.

Hagemann, E. and Pestel, R. (1987) Agriculture as a component of China's modernization strategy, in *Learning from China – Development and Environment in Third World Countries*, Glaeser, B. ed., Allen & Unwin, London, pp162–172.

Hay, R.W. (1979) Statistics on food and agriculture in China, *Food Policy*, November 1979, pp295–299.

Hazell, P.B.R. and Norton, R.D. (1986) *Mathematical Programming for Economic Analysis in Agriculture*, Macmillan Publishing Company, New York.

He Changmao (1985) The Prospects of China's Livestock Industry and Issues to be Solved Currently, in *On the Development Strategy of Chinese Countryside*, Lu Wen, Niu Ruo Feng & Hu Yi Zun, eds, Chinese Press of Agricultural Science and Technology.

He Kang (1984) Deepening the Reform and Readjusting the Production Structure to Simulate an Extensive Development of the Rural Commodity Economy, *Agrotechnical Economy*, No. 12, 1984, pp1–5.

He Kang (1986) Carrying on the Reform to Promote Agricultural Development in a Steady and Coordinated Way, *Agrotechnical Economy*, No. 2, 1986, pp1–7.

Heady, E.O., Baker, C.B., Diesslin, H.G., Kehrberg, E. and Staniforth, S. eds. (1961) *Agricultural Supply Functions: Estimating Techniques and their Interpretations*, Iowa State University Press, USA.

Higgins, B. (1959) *Economic Development*, W.W. Norton and Co., New York.

Ho Ping-ti (1959) *Studies on the Population of China, 1368–1955*, Cambridge: Harvard University Press.

Hoeffding, O. (1959) State Planning and Forced Industrialization, *Problems of Communism* (Rand Cooperation), Vol. 8, Nov–Dec., pp38–46.

Huan Xiang (1985) On Reform of Chinese Economic Structure, *Beijing Review*, No. 20, May 20, 1985.

Huang Deimin (1985) Strategy for the Development of Crop Seeding, in *On the Development Strategy of Chinese Countryside*, Lu Wen, Niu Ruo Feng, & Hu Yi Zun, eds, Chinese Press of Agricultural Science and Technology, pp502–511.

Hussian, A. and Tribe, K. (1981) Russian Marxism and the Peasantry 1861–1930, In *Marxism and the Agrarian Question*, Vol. 2 Macmillan.

Hussian, A. (1986) Science and Technology in the Chinese Countryside. A paper prepared for the conference on China's New Technological Revolution, May 9–11, 1986, Harvard.

International Trade Research Institute of the Ministry of Foreign Economic Relations and Trade (1984) Reviews of International Market Conditions for Grain and Cash Crops and Prospects on China's Future Trade Stance, in *Study of the Development of Grain and Cash Crops Production in China*, Vol. 1. Edited by the Grain and Cash Crops Development Research Group, CAAS, pp145–60.

Investigation Group of the Chinese Academy of Agricultural Science (1985) On Issues concerning the Current Economic Restructuring in China's Rural Area, *Agrotechnical Economy*, No. 11, 1985, pp1–4.

Ishikama, S. (1965) *National Income and Capital Formation: An Examination of Official Statistics*, Institute of Asia Economic Affairs, Tokyo.

Ji Cai (1984) On How to Modify the Grain Subsidy Policy, *Economics of Finance and Trade*, December 1984, pp49–50.

Johnston, B.F. and Kilby, P. (1975) *Agriculture and Structural Transformation: Economic Strategies in Late–Developing Countries*, New York: Oxford University Press.

Keng Chien-hua (1981) How Can We Calculate Whether the Rate of Increase in Budgetary Revenue is Appropriate?, Tsái-Cheng, 4:17–19.

Klatt, W. (1983) The Staff of Life: Living Standards in China, 1977–81, *China Quarterly*, No. 93, pp17–50.

Koutsoyiannis, A. (1984) *Theory of Econometrics*, Second Edition, Macmillan Publishers Ltd., London and Basingstoke.

Kraus, W. (1982) *Economic Development and Social Change in the People's Republic of China*, Translated by Holz, E.M., Springer-Verlag. New York-Heidelberg-Berlin.

Kraus, W. (1984) Economic Development and Social Change in the Projections for 1985, 1990 and 2000, *Asian Survey* Vol. 24, Part 12, pp1247–1274.

Kueh, Y.Y. (1984) China's New Agricultural Policy Program: Major Economic Consequences, 1979–1983, *Journal of Comparative Economics*, Vol. 8, No. 4, pp353–375.

Kuznets, S. (1964) Economic Growth and the Contribution of Agriculture: Notes on Measurements, in *Agriculture in Economic Development*, Eicher and Witt eds, McGraw-Hill, New York.

Lardy, N.R. (1976) Economic Planning and Income Distribution in China, *Current Scene*, Vol. XIV, No. 11, pp1–12.

Lardy, N.R. (1978) *Economic Growth and Distribution in China*, Cambridge University Press, Cambridge.

Lardy, N.R. (1983) *Agriculture in China's Modern Economic Development*, Cambridge University Press, Cambridge.

Leeming, F. (1985) *Rural China Today*, Longman, London.

Leibenstein, H. (1957) *Economic Backwardness and Economic Growth*, John Wiley, New York.

Li Deping (1980) Cotton Price and Cotton Production in China since Liberation, *Beijing University Bulletin*, No. 4, 1980, pp2–9, 27.

Liu Hsien-Kao (1977) Agricultural Production Planning Tables, in Lardy, N.R. ed., Chinese Economic Planning, translations from Chihua Ching-Chi, M.E. Sharpe, Inc. Danson, pp179–189.

Liu Suinien (1980) The Proposal and Implementation of the Eight Characters of Readjustment, Consolidation, Filling-out, and Raising Standards, in *Party History Research (Tang-shih yen-Chiu)*, No. 6, pp21–33.

Lu Liangshu (1984) Agricultural Scientific Research Work is to be Geared to the Needs of the Economic Construction, *Agrotechnical Economy*, No. 9, September 1984, pp1–6.

Luo Chuhua (1983) The Development of the Agricultural Responsibility System and Questions Raised in the Course of its Application, in *Problems of Agricultural Economics (Nongye Jingji Wenti)*, March 1983, pp3–9.

Luo Manxian (1985) *Economic Changes in Rural China*, China Studies Series, New World Press, Beijing, China.

Mah Feng-Hwa (1971) Why China imports Wheat, *China Quarterly*, No. 45, Jan-March 1971, p128.

Mai Huangquan (1984) On the future development of China's rice production, in Study of the Development of Grain and Cash Crops Production in China,Vol. 2, Edited by the Grain and Cash Crops Development Research Group, CAAS, March 1984, pp1–11.

Mao Zedong (1942) Economic and Financial Problems in the Anti-Japanese War, in *Selected Works of Mao Zedong*, Vol. 3, pp111–116, (1967) Foreign Languages Press, Peking.

Mao Zedong (1943) On Coalition Government, in *Selected Works of Mao Zedong*, Vol. 3, pp205–270, (1967) Foreign Languages Press, Peking.

Mao Zedong (1955) On the Question of Agricultural Cooperativization, *New China Monthly (Hsin-hua Yeh-Pao)*, Vol. 73, No.11, pp1–8.

Mao Zedong (1956) On Ten Major Relationships, in *Selected Works of Mao Zedong*, (1986) People's Press, Beijing, China.

Ministry of Textile Industry Research Office (1981) China's Textile Industry, in *Annual Report of China (1981)*, Xue Mugiao ed. Overseas Chinese Language Edition, Hong Kong Modernization Company, Hong Kong.

Murphey, R. (1982) Natural Resources and Factor Endowments, in *The Chinese Agricultural Economy*, Barker, Sinha and Rose eds, Westview Press/Croom Helm.

Nerlove, M. (1956) Estimates of Supply of Selected Agricultural Commodities, *J. Farm. Econ.* 38:496–509.

Nerlove, M. (1958) *The Dynamics of Supply: Estimation of Farmer's Response to Price*, Johns Hopkins University Press, Baltimore.

Nerlove, M. (1979) The Dynamics of Supply: Retrospect and Prospect, *Amer. J. Agr. Econ.* 61: 874–888.

Nerlove, M. and Bachman, K.L. (1960) The analysis of changes in agricultural supply: problems and approaches, *J. Farm. Econ.* 42(3): 531–554.

Nicholls, W.H. (1964) The Place of Agriculture in Economic Development, in *Agriculture in Economic Development*, Eicher and Witt eds, McGraw-Hill, New York.

Niu Luohwen (1985) Technological Improvement and Agricultural Modernization, in *On the Development Strategy of Chinese Countryside*, Lu Wen, Niu Ruo Feng & Hu Yi Zun eds, Chinese Press of Agricultural Science and Technology, pp315–338.

Nolan, P. (1976) Collectivization in China: Some Comparisons with the USSR, *Journal of Peasant Studies*, Vol. 3, No. 2, pp192–220.

Norusis, M.J. (1985) *Advanced Statistics Guide -SPSS* McGraw-Hill, New York.

OECD (Organisation for Economic Co-operation and Development) (1985) *Agriculture in China: Prospects for Production and Trade*, Paris.

Parish, W.L. and Whyte, M.K. (1978) *Village and Family in Contemporary China*, University of Chicago Press, Chicago.

People's Daily (1977) Revolution plus production can solve the problems of food, October 20, 1977, p2.

People's Daily (1986a) The CCCP and State Council's instruction on rural development, Overseas Edition, Feb. 23, 1986, p3.

People's Daily (1986b) Encouraging outcome has come out from the construction of the fifty commercial grain production bases, Overseas Edition, March 8, 1986, p1.

People's Daily (1987a) China's population growth is rising again, Overseas Edition, Feb 17, 1987, p1.

People's Daily (1987b) Significant improvement in Chinese people's life standard, Overseas Edition, Oct. 17, 1987, p1.

People's Daily (1987c) It is imperative to protect the arable land, Overseas Edition, May 18, 1987, p1.

People's Daily (1987d) The implementation of 'Bumper Harvest Programme' is progressing smoothly, Overseas Edition, June 20, 1987, p4.

People's Daily (1987e) The new rural banking system is under establishment, Overseas Edition, June 29, 1987, p1.

People's Daily (1987f) The total population in China has reached a new record of 1.07 billion persons, Overseas Edition, Nov. 12, 1987, p1.

People's Daily (1987g) Data on the new population sampling survey is published by the SSB, Overseas Edition, Nov. 12, 1987, p1.

People's Daily (1987i) Zhao talks about the main themes of China's restructuring to the Japanese, Overseas Edition, Nov. 14, 1987, p1.

People's Handbook, (1956) Temporary Measures for Rationed Supply of Grain in Cities and Townships and Temporary Measures for Planned Purchasing and Marketing of Grain in Rural Areas, Tientsin: Ta Kung Pao, p488.

Perkins, D.H. (1966) *Market Control and Planning in Communist China*, Cambridge Harvard University Press.

Perkins, D.H. (1975) Constraints influencing China's agricultural performance, in *China: A Reassessment of the Economy*, Joint Economic Committee, Congress of the US, Washington, DC.

Qiao Longzhang (1985) Study of the Rôle played by the Law of Value from the Experiences of Agricultural and Sideline Development, *Study and Research*, No. 1, 1985, pp25–26.

Qu Haibo (1987) On the current trend of population growth in China, *People's Daily*, Overseas Edition, July 30, 1987, p8.

Rawski, T.G. (1979) *Economic Growth and Employment in China*, published for the World Bank, New York; Oxford University Press.

Riskin, Carl (1987) *China's Political Economy; The Quest for Development since 1949*, Oxford University Press.

Roningen, V.O., Dixit, P.M. and Seeley, R. (1988) *Agricultural outlook for the Year 2000; Some Alternatives*. A research paper presented at the XX International Conference of Agricultural Economists, Buenos Aires, Argentina, August 26 -September 2, 1988.

Sanderson, B.A., Quilkey, J.J. and Freebairn, J.W. (1980) Supply Response of Australian Wheat Growers, *Aus. J. Agri. Econ.*, Vol. 24, No. 2, pp 129–140.

Schran, P. (1982) Agriculture in the Four Modernizations, in *Agricultural Development in China, Japan and Korea*, Hou and Yu eds, Academia Sinina, Taiwan.

Schultz, T.W. (1964) *Transforming Traditional Agriculture*, Yale University Press, New Haven.

Seini, A.H. (1984) "An Economic Study of Production, Supply and Processing of Seed Cotton in Ghana", a Ph.D. thesis of Wye College, University of London.

Shen, T.H. (1951) *Agricultural Resources of China*, Cornell University Press.

Shen Zhiping (1984) Analysis of the current dietary composition in China and predicting its future development trend, in *Study of the Development of Grain and Cash Crop Production in China*, Vol. 1. Edited by the Grain and Cash Crops Development Research Group, CAAS, March 1984, pp115-122.

Simon, András (1980) *An Econometric Model of the Hungarian Economy*, Contributed paper to the 4th World Congress of the Econometric Society, Institute for Economic and Market Research, Aix-en-Provence, 28.8 –3.9.1980.

Smil, V. (1981) China's food: availability, requirements, composition, prospects. *Food Policy*, May 1981, pp67–77.

Song Jian, Tian Xue Yuan, Yu Jung Yuan, & Li Guang Yuan, (1982) *Population Forecasting and Population Control*, The People's Press, Beijing, China.

Staniforth, S.D. and Diesslin, H.G. (1961) Summary and Conclusions, in Heady, E.O., Baker, C.B., Diesslin, H.G., Kehrberg, E., and Staniforth, S. eds, *Agricultural Supply Functions: Estimating Techniques and Interpretations*, pp293–302.

State Statistical Bureau ed. (1983) *Chinese Statistical Yearbook, 1983 (Zhongguo Tongji Nianjian, 1983)*, Chinese Statistical Press, Beijing, China.

State Statistical Bureau ed. (1984) *Statistical Information China's Trade and Prices, 1952–1983 (Zhongguo Caimao Wujia Tongji Ziliao 1952–1983)*, Chinese Statistical Press, Beijing, China.

State Statistical Bureau ed. (1985) *Chinese Statistical Yearbook, 1985, (Zhongguo Tongji Nianjian, 1985)*, Chinese Statistical Press, Beijing, China

State Statistical Bureau ed. (1986) *Interpretations of the main statistic items in Commerce*, December 1986.

State Statistical Bureau ed. (1987) *Chinese Statistical Yearbook, 1987, (Zhongguo Tongji Nianjian, 1987)*, Chinese Statistical Press, Beijing, China.

Stavis, B. (1982) Rural Institutions in China, in *The Chinese Agricultural Economy*, Barker, Sinha and Rose eds, Westview Press/Croom Helm.

Stone, Bruce (1982) The Use of Agricultural Statistics: Some National Aggregate Examples and Current State of the Art, in *The Chinese Agricultural Economy*, Barker,Sinha and Rose eds, Westview Press/Croom Helm.

Stone, Bruce (1989) Chinese Wheat Production and Technical Change, in *The Wheat Revolution Revisited: Recent Trends and Future Challenges*, ed. by CIMMYT, Mexico, pp17–21.

Sung Jian, Tian Xue Yuan, Yu Jing Yuan and Li Guang Yuan (1982) *Population Forecasting and Population Control*, The People's Press, Beijing, China.

Sun Wei-tzu (1958) Principles for Organizing Grain Circulation Planning, in *Planned Economy (Chi-hua Ching-Chi)*, 1958, No. 2, pp24–27).

Surls, F.M. (1982) Foreign Trade and China's Agriculture, in *The Chinese Agricultural Economy*, Barker, Sinha and Rose eds (1982), Westview Press/Croom Helm, pp183–198.

Tang, A.M. (1967) Agriculture in the Industrialization of Communist China and the Soviet Union, *J. of Farm Econ.* (December 1967): pp1118–1134.

Tang, A.M. (1968) Policy and Performance in Agriculture, in *Economic Trends in Communist China*, Eckstein, Galenson and Liu eds, Edinburgh University Press, Edinburgh.

Tang, A.M. and Stone, B. (1980) Food Productions in the People's Republic of China, *International Food Policy Research Institute Research Report 15*, May 1980.

Teng Tzu–hui (1955) *New China Monthly (Hsin-hua Yeh-pao)*, pp9–13, November 1955.

The Committee for the World Atlas of Agriculture ed. (1973) *World Atlas of Agriculture, Vol. 2, Asia and Oceania*, Instituto Geografico De Agostini-Novara.

The Economist (1987a) Turning grain into pigs, January 24, 1987, p40.

The Economist (1987b), Baby Trouble, May 2, 1987, p48.

Thomas, W.J. ed. (1972) *The demand for food: An exercise in household budget analysis*, Manchester University Press.

Timmer, C.P. and Jones, J.R. (1986) China: An Enigma in the World Grain Trade, in *East -West Agricultural Trade*, Jones, J.R. ed.Westview Press, Boulder and London, pp153–180.

Timmer, C.P. (1976) Food Policy in China, *Food Research Institute Studies* Vol. XV, No. 1, pp53–70.

TJNJ, *see* State Statistical Bureau (SSB).

Tomek, W.G. (1972) Distributed Log Models of Cotton Acreage Response: A Further Result, *Amer. J. Agr. Econ.* 54: pp108–110.

Tomek, W.G. and Robinson, K.L. (1981) *Agricultural Product Prices*, Cornell University Press, Ithaca and London.

Tong Da lin and Bao Tong (1978) Some Views on Agricultural Modernization, *Peoples Daily*, December 8, 1978.

Walker, K.R. (1965) *Planning in Chinese Agriculture*, Frank Gass Company Ltd., London and Edinburgh.

Walker, K.R. (1968) Organisation of Agricultural Production, in *Economic Trends in Communist China*, Eckstein, Galenson and Liu eds, Edinburgh University Press.

Walker, K.R. (1977) Grain Self-sufficiency in North China, 1953–75, *China Quarterly*, No. 71.

Walker, K.R. (1982) Interpreting Chinese Grain Statistics, *China Quarterly*, No. 92, pp575–588.

Walker, K.R. (1984) *Food Grain Procurement and Consumption in China*, Cambridge University Press, Cambridge.

Walker, N. and Moneypenny, R. (1976) Linear Programming as a Tool for Agricultural Sector Analysis, *Review of Marketing and Agricultural Economics*, Vol.44, No.4.

Wang Hsiang-Ch'un, Chiang Hsing-ei and Chen k'un-hsui (1965) Problems of Controlling Plans for Agricultural Production in China, in *Economic Research (Ching-chi yen-chin)*, 3: pp33–9.

Wang Keng-chin (1959) An Appreciation of Several Points in Agricultural Planning Work, in *Planning and Statistics (Chi-hua y t'ung-chi)*, 14: pp15–21.

Wang Yinwei (1985) On Controlling China's Rural Population Growth, in *On the Development Strategy of Chinese Countryside*, Lu Wen, Niu Ruo Feng & Hu Yi Zun eds, Chinese Press of Agricultural Science and Technology.

Wiles, P.J.D. (1962) *The Political Economy of Communism*, Basil Blackwell, Oxford.

White, G. (1982) Introduction: The New Course in Chinese Development Strategy: Context, Problems and Prospects, in *China's New Development Strategy*, Gray and White eds, Academic Press, pp1–16.

Wong, John (1973) *Land Reform in the People's Republic of China*, Praeger Publisher, New York-Washington-London.

Wong, John (1979) Rice Exports: A New Dimension in China's Economic Relations with Southeast Asia, *J. Southeast Asian Studies*, Vol. 10, Part 2, pp451–469.

Wong, John (1980) China's Wheat Import Programme, *Food Policy*, November, pp318–330.

World Bank (1985a) *China: Long-Term Development Issues and Options*, Johns Hopkins University Press, Baltimore and London.

World Bank (1985b) China: Agriculture to the year 2000, Annex 2 to *CHINA Long-Term Development Issues and Options*, A World Country Economic Report, Johns Hopkins University Press, Baltimore and London.

Wu shih (1957) A Discussion of the Grain Question in China during the Transition Period, in *New China Semimonthly (Hsin-hua pan-yeh kan)*, Vol. 108, No. 10, pp104–109.

Wu Shuo (1985) Review of recent agricultural production and suggestive solutions for the future development, in *Study of the Development of Grain and Cash Crop Production in China*, Vol. 4, Edited by the Grain and Cash Crops Development Research Group, CAAS, June 1985, pp108–16.

Wu Xiang (1983) Combined Contract Responsibility System and Agricultural Planning, in *Problems of Agricultural Economics (Nugh-yeh Ching-chi wen-t'i)*, Feb. 1983, pp3–10.

Xiao Chuoji (1980) The Law of Price Movement in China, *Social Sciences in China*, No. 4, 1980, pp44-59.

Xue Mugiao (1980) *The Rural Economy in Rural China*, Agricultural Publishing House, Beijing, China.

Xue Mugiao (1985) Rural Industry Advances Amidst Problems, *Beijing Review*, Vol. 28, No. 50, pp18–21.

Xue Mugiao (1987) Get rid of dogmatism and the ossified traditional model, *People's Daily* Overseas Edition, Dec. 8, 1987, p2.

Yan Yubai (1984) Projection of the future developments of China's wheat production and consumption, in *Study of the Development of Grain and Cash Crop Production in China*, Vol. 2. Edited by the Grain and Cash Crops Development Research Group, CAAS, March 1984, pp12–26.

Yang Jianbai and Li Shucheng (1980) On the Historical Experience of the Relations between Agriculture, Light Industry and Heavy Industry in China, *Chinese Social Science*, No. 3, 1980, pp19–44.

Yen Shao-ch'un, (1977) Agricultural Production Planning, in *Chinese Economic Planning*, (Translations from Chi-hua Ching-Chi) Lardy, N.R. ed., M.E. Sharpe, Inc. Dawson, pp24–36.

Youngson, A.J. (1959) *Possibilities of Economic Progress*, Cambridge University Press, London.

Zhan Wu (1980) A Correct Decision on the Development of Agriculture, *Problems of Agricultural Economics*, 1980, No. 8, pp2–8.

Zhang Di (1984) Approaching on the development of China's soybean production, in *Study of the Development of Grain and Cash Crop Production in China*, Vol. 2. Edited by the Grain and Cash Crops Development Research Group, CAAS, March, 1984, pp48–60.

Zhang Tong (1984) Estimations for China's future food consumption and nutritional intake conditions, in *Study of the Development of Grain and Cash Crop Production in China*, Vol. 1. Edited by the Grain and Cash Crops Development Research Group, CAAS, March 1984, pp118–122.

Zhang Tong, (1985) Analysis of the World Cereal Development Conditions and Study of the Principle for China's Grain Development in *On the Development Strategy of Chinese Countryside*, Lu Wen, Niu Ruo Feng, & Hu Yi Zun, eds, Chinese Press of Agricultural Science and Technology, pp490–501.

Zhang Yu'an (1989) Industries racing up ahead of agriculture, *China Daily*, March 1, 1989, p1.

Zhao Ziyang (1985) Why Relax Agricultural Price Controls?, *Beijing Review*, Vol. 28, No. 7, pp16–18, 29.

Zhao Changhu and Ming Haijun (1985) Grain production and the Law of Value, *Agrotechnical Economy*, No.4, 1985, pp38–49.

Zhong Guangjun (1985) Some Issues to be Solved in the Grain Production, *Agrotechnical Economy* No.7, 1985, pp18–21.

Zhou Sinning ed. (1984) *Comprehensive Agricultural Zoning of Jianou County (Jianou Xian Zonghe Nongye Quhua)*, Agricultural Zoning Office of Jianou County.

Index